Longman exam practice kits

A-Level Geography

John Smith

LONGMAN

Series Editors
Geoff Black and Stuart Wall

Titles available

A-Level
Biology
British and European Modern History
Business Studies
Chemistry
Economics
French
German
Geography
Mathematics
Physics
Psychology
Sociology

Addison Wesley Longman Ltd,
Edinburgh Gate, Harlow,
Essex CM20 2JE, England
and Associated Companies throughout the World.

First published 1997

ISBN 0582-31254-X

British Library Cataloguing-in-Publication Data
A catalogue record for this book is available from the British Library.

Set by 30 in 10.5/12.5 Baskerville MT
Produced by Longman Singapore Publishers, pte
Printed in Singapore

Acknowledgements

The authors and publishers are grateful to the following examination boards for permission to
reproduce their questions:

▶ Associated Examining Board (AEB)

▶ EDEXCEL Foundation (London)

▶ Northern Examinations and Assessment Board (NEAB)

▶ Welsh Joint Education Committee (WJEC)

▶ University of Cambridge Local Examinations Syndicate (Oxford)*

*UODLE (Oxford) material is reproduced by permission of the University of Cambridge Local
Examinations Syndicate.

All answers to questions supplied by the examination boards are entirely the responsibility of the
authors and have neither been provided nor approved by the examining boards.

Contents

How to use this book iv

Part I   Preparing for the examination 1

Part II   Topic areas, summaries and questions 11

Chapter 1   Ecosystems 12

Chapter 2   River systems 18

Chapter 3   Plate tectonics 23

Chapter 4   Meteorology and climate 30

Chapter 5   Population and resources 41

Chapter 6   Settlement 47

Chapter 7   Employment 54

Chapter 8   World and regional development 61

Part III   All the answers 67

Chapter 1   Solutions: Ecosystems 68

Chapter 2   Solutions: River systems 74

Chapter 3   Solutions: Plate tectonics 79

Chapter 4   Solutions: Meteorology and climate 84

Chapter 5   Solutions: Population and resources 90

Chapter 6   Solutions: Settlement 96

Chapter 7   Solutions: Employment 102

Chapter 8   Solutions: World and regional development 109

Part IV   Timed practice paper, answers and mark schemes 115

Timed practice paper 116

Answers and mark scheme 120

How to use this book

This book sets out to help you to achieve your best possible grade in the A-level geography examination. It covers the topics which are most common across all the A-level syllabuses from the different examination boards. However, no single book can cover all the topics from all the boards, and you must make sure that you concentrate on the sections of the book which are relevant to your own syllabus. You really need to obtain your own copy of your syllabus from the board, or at least get hold of a copy of the subject content. Your subject teacher is probably the best person to supply this.

The book is arranged in four parts.

Part 1   Preparing for the examination

Here we consider some useful techniques that you might use in preparing for, and sitting, the examination. A revision planner is provided to help you structure your work in the period leading up to the examination. The different types of question you might face are considered, along with the techniques you will need to use in answering each type of question successfully. It is essential that you interpret each question correctly, so a 'glossary' has been provided. This gives the key **command words** such as define, assess, etc. which actually tell you what to do in each question.

Part II   Topic areas, summaries and questions

Here eight topics have been identified – four on physical geography and four on human geography – although you will notice there are many overlaps between the different sections. For each topic you will find:

1 **Revision tips** Giving guidance on revising that particular area.
2 **Topic outlines** Briefly summarising the key knowledge and understanding in that area of the syllabus.
3 **Revision activities** To encourage you in active revision there are exercises which help you organise and check your learning.
4 **Practice questions** Which provide examples of the types of task you will be asked to do in your examination.

It is essential that you try the questions before you check the answers provided in Part III of the book. Writing practice examination answers is the best possible way to prepare for the experience ahead – of writing real examination answers.

Part III   All the answers

Here you will find answers to the activities and practice questions set on each topic area. Some of these answers will be actual student answers, with comments from an examiner, which should help you to identify strengths and weaknesses in the answer. Other questions will be provided with outline answers from examiners, to show what they were looking for in the question.

You will also find some extracts from mark schemes, to show how the examiners distinguish between answers at low, middle and high levels of attainment.

Part IV   Timed practice paper, answers and mark schemes

Here you will find questions and parts of papers which will provide you with practice in timing yourself under exam-type conditions in preparation for the examination. Outline answers and a breakdown of the marks are provided to these questions so that you can check your own performance.

part I

Preparing for the examination

Preparing for the examination

PLANNING YOUR REVISION

▶ Start revising in good time. Last minute revision does work for some people but, on the whole, it is risky.

▶ Aim to build up an organised structure of knowledge throughout your revision period. Start by developing a general outline of a topic; then add in detail every time that you come back to that topic. It is much easier to learn the detail when you are fitting it into a clear context.

▶ Plan your time carefully. Make sure that you have enough 'revision slots' to cover *all* the topics, in *all* the subjects you are taking.

▶ Be clear about what you have to learn. Make a list of all the topics you have covered in the course. Be *absolutely sure* which of these are compulsory, and which are optional in the examination.

▶ Divide your topics into:
   – **essential**, which you will certainly have to do in the examination, or which you will rely on in the optional sections
   – **important**, which you will probably have to do, either in a compulsory or an optional section
   – **useful**, which will allow you extra choice in the optional sections, if something goes wrong with some of the other questions

You must make sure that the topics in the first two groups are learnt thoroughly, and hope to do as many as possible from the third group – because accidents do happen!

▶ Be prepared to adapt your timetable as you go along.

▶ Build rest periods and rewards into your timetable. Use your rest periods to keep your mind fresh. Keep as fit as possible, and as fresh as possible. Fresh air, exercise, and good food are vitally important during revision periods. Rewards can be simple little things like:
   – sticking gold stars on your 'revise guide revision planner' when you have achieved a task
   – allowing yourself to watch that TV programme, with a clear conscience, after you have finished a task
   – taking the dog out for a walk
   – ringing your boy/girl friend, as long as you know you will not be disturbing his/her revision
   – having an ice cream/chocolate biscuit/apple, when you have done an hour of hard work
   – or whatever gives you something to look forward to, when you have achieved a certain goal

▶ Revise actively. This means doing something more than just reading and rereading. You may think you have understood what you have read; but this is not the same as being able to reproduce and develop these ideas in an examination. At the very least you should rework your notes, highlighting key points, making lists of key ideas and facts, developing flow diagrams or 'mind maps' of ideas. Particularly important in geography is the process of learning, and then redrawing, maps and diagrams. These are very efficient ways of summarising and organising a lot of information about a place, in a very small space.

▶ Practise reproducing information without the aid of notes, especially by completing practice examination questions. You could also try reciting model answers while you are lying in the bath. As long as you can still concentrate on your road skills, analyse the physical geography of the area when you are out on your bike. Risk social ridicule by describing the problems of urban transport systems to your pals when you are on the bus to school. Really impress your parents by turning down the sound on the TV weather forecast, and give your own forecast, interpreting the satellite images and symbols on the charts. Geography is all around you; live it, and use it.

▶ Work hard! There are very few natural geniuses who can learn information without effort; but remember, quality revision is more important than quantity.

▶ Stay friends with the people you live with. They want you to do well, and they probably care about you – even if they sometimes show it in a funny way. Reassure them, by showing them you are doing your best; and accept their offers of help – but make sure that it is what you need and not interference. You might organise them to:
  – bring you tea, or cold drinks when you need them
  – test you (but be sure parents only test you on your up-to-date syllabus, not on what they seem to remember learning 20 or more years ago)
  – make sure that you can relax when you have finished a revision session
  – make sure that your workplace is quiet, calm and free from interruptions
  – listen to your obsessive repetition of lists of facts

But remember, you will only keep them on your side if you show them some understanding too. Let them see that you are doing the best you possibly can.

USING EXAMINATION QUESTIONS IN REVISION

After you have revised a topic it is essential that you carefully consider how to *use* your knowledge to answer the type of questions you will face in your examination. Try to answer the questions you will find in Part II of this book, before looking at the answers in Part III. This makes your revision into an *active* process, and it has many advantages:

▶ You can check whether you really understand what you have tried to revise.

▶ It reveals any gaps in your knowledge. This means that you can go back and revise more thoroughly. You may add more to your summary notes, or you may have to revise a new sub-topic. It is far better to reveal the gaps now, and plug them, rather than finding out on the day of the examination.

▶ You will start to become more familiar with 'exam-speak', particularly the type of command words used by examiners (see the glossary on p. 9).

▶ You will develop an understanding of what examiners are looking for, by reading the outline answers in Part III.

▶ You will build up a 'bank' of answers. It is very reassuring to see a question in an examination and realise that you have already written an answer to a very similar question. This gives you a head start in the examination – as long as you can adapt your answer to fit the question asked.

▶ As the examinations get nearer you should practice answering questions, and even whole papers, under timed 'examination conditions'. Part IV of the book should be useful here.

Types of examination questions

Data response and practical questions

In these questions the examiner presents you with at least one, but often more, pieces of data. These will usually be unfamiliar to you, and you have to respond in a logical, structured, geographical way. The data may be in the form of maps (including OS maps), photographs, tables of statistics, sketches, graphs, remote sensing images, text, newspaper articles, etc. In fact examiners often try to provide the biggest possible range and variety of sources, to test your ability to apply the skills you have learnt during your course.

You will be tested on some, or all, of the following:

▶ **Data collection** You will need to know about sampling methods, and ways of collecting and recording in the field. You may well be asked about the reliability of different methods of collection. It is often useful to refer to your own fieldwork experiences, and it may be very useful to refer to particular problems that you encountered in your work.

▶ **Sources of data** Some syllabuses ask very specific questions about the sources of data that you would need to research certain subjects. For these you need to name one or two sources, and then describe one of them briefly.

▶ **Data presentation** You may be asked to draw graphs or maps, but questions are more likely to ask for discussion of the advantages and disadvantages of different ways of presenting data. You might be asked to criticise one or more ways of showing information. If you are given a map or a graph and asked to criticise the technique, you must comment on both the strengths *and weaknesses* of the presentation. Give your opinions clearly, but you must support them with evidence.

▶ **Analysis** Having selected and presented the data, you need to use it. You may be asked to use one or more statistical techniques, or to comment on the relevance of techniques. As this is a geography examination you will not be asked to do complicated sums, but rather to write about how, where and when to use the techniques, or how to explain the results when you have obtained them. In this section you should be looking for patterns or trends in the data. Then, having seen patterns, you should note where anomalies occur, or where the pattern is not followed.

▶ **Explanation** or **conclusion** Finally you will be asked to show that you recognise the reasons for the facts and distributions shown by the data. It is important that you recognise relationships between the different pieces of information that you have been looking at, because it is here that geographers show the highest levels of skill and understanding.

You should find that your answer to the final part of such questions contains many words and phrases such as 'because', 'therefore', 'and so', 'compared with', 'this relates to', 'on the other hand', and so on. Each of these phrases is a way of pointing out a link or a connection between different pieces of data. Apply the full range of skills that you have gained on your course to the new data.

Be careful to make sure that you do not just write about the data without showing evidence that you can fit it into the whole body of geographical studies. The data may be new to you; but the ideas that help to explain it should be very familiar, and you will not gain high marks without linking the new information to that body of geographical knowledge and understanding.

Shorter, structured questions

In these questions the examiners are looking for you to show that you understand some of the basic, key ideas, principles and theories of geography. Most structured answer papers have several questions, usually compulsory, which test parts of the core syllabus. They are often supported by simple resource material.

You must be especially disciplined when answering these papers. You must always work within a time discipline and often within a space discipline. If you are given a number of lines to write on you ought to use all or most of the space, but you should not go beyond it – except in an emergency.

Work out how much time you have for the whole paper; then divide your time between the questions. This may leave you only 10 or 15 minutes for each question, but you must remember that the mark scheme will have been designed so that you can get the full marks within that time limit. It is most unlikely that you will double your marks for a question if you spend twice as long on your answer. In fact it is policy with some examination boards *not* to mark any part of an answer which goes beyond the space allowed in the answer book. You must be disciplined in your approach.

▶ If there is only 1 mark available for an answer you probably only need to write one sentence, one phrase, or maybe only one word.

▶ If there are 2 marks it means that you are being asked for an idea followed by some kind of simple elaboration, explanation or example.

▶ For 3–5 marks you need to write a paragraph, or a series of points, but be concise. Try to make sure that each sentence or point moves logically and moves from simple ideas to more developed, complex ones.

The first part of a structured question often asks for recall of a key piece of information, or a definition. Your revision should take this into account. Make sure that you know your definitions, and can write them out quickly and clearly, without fuss. The later parts usually ask you to apply your knowledge, or expand on your original idea. This may well take longer than the first part, but be concise. Think about your answer before you start to write. You should know how each sentence is going to end *before* you start to write it.

If you come across a question that you cannot do immediately, do not despair. Check that you are not being given vital clues in the resource material, or in the rest of that question. If, after consideration, you still cannot do it go on to the next question; but mark the question you have missed with a star, and come back to it later. You may have remembered something vital which gives you the key. Whatever happens, though, try to write something. You will not lose marks for writing a wrong answer. Wrong answers are ignored, but a sensible 'educated guess' just might be lucky, and gain an extra mark or two!

Essay questions

Twenty years ago A-level geography examinations were almost entirely based on writing essays. Now you need a far wider range of skills, as shown above, but essays are still an essential part of all syllabuses. Unfortunately they are often badly done. Candidates do not have to do much essay writing at GCSE, and they often seem to lack experience in practising A-level essay writing. This book offers you a lot of advice on essay technique, but only you can do the practice.

You almost always have a choice of essay questions. Then you usually have about 45 minutes for each essay. However, there is a lot to do before you start to write.

1 Be sure you know how many essays you must write, and which sections you must choose from. Every year some candidates destroy their chances by making *rubric errors* and doing the wrong number or type of questions. Don't let it be you!

2 Choose logically. Look for essays on topics that you are very familiar with. It is particularly good to pick a title which is similar to an essay that you have done before.

3 Make sure that you can do all parts of your chosen question. It is no good answering part i well, for 7 marks, if you cannot do part ii properly, when it is worth 18. You must be certain that, if the question asks for 'references to one or more case studies', you know relevant examples in detail. Do not be seduced by attractive source material. It is silly to do a question because it has a dramatic, interesting photo, or an amusing cartoon, when you should realise that you have not revised the topic thoroughly.

4 Analyse the question. Highlight, or underline, the key words. Pay particular attention to 'command words' (see the glossary on p. 9), but also note words such as 'examples', 'more' and 'less developed', 'change', and so on. Be sure that you know exactly what the examiner wants from you.

5 Plan your answer. If you have 45 minutes altogether it is worth spending between 5 and 7 minutes on the plan. Planning an essay is like planning a journey. Remember, when good geographers have planned a journey they keep referring back to the map to check they are still on the right route. While you are writing your essay keep referring back to the plan to make sure you are still going the right way!

Journey planning	Essay planning
• Know where you are starting from. • Know where you want to go. • Plan which roads will get you there quickly and directly. • Note down some of the landmarks which will guide you. • Think about other interesting places you *could* visit, if you have time and energy. • How long will the journey take? When will you arrive?	• Plan the introduction. • Know what conclusion you are aiming to reach. • Note down the logical sequence of ideas which you will develop as you go through the essay. • Think about which case studies and key facts you will use to illustrate your ideas. • What other relevant, linked ideas *could* you bring in to expand on the theme, if you have time or space. • How long can you spend on each part of the essay, given your time constraints?

The **introduction** is often the most important part of the whole essay. Should you go for a dramatic, 'headline', brilliant start, which adopts a very clear point of view; or adopt a more businesslike, down to earth, sober style?

Sometimes you might be able to write a really original opening, to grab the examiner's attention, and give a flavour of the drama to come. If you can do this well then you will gain a lot of credit for it; but it is difficult to do, and rather risky, unless you know the topic really well, and can support your opinions with very detailed facts and ideas.

The more reliable introduction does some or all of the following:

► defines the terms that are used in the question
► outlines the factors that affect the topic under consideration
► outlines the structure you will follow in your answer
► introduces the places and examples you will discuss later
► makes a clear statement of a question which you can answer later, in your conclusion

The **development section** deals with each of the points in your plan in turn. Try to make sure that each sentence introduces something new. This could be:

► a new idea
► an elaboration of an idea mentioned before
► a link between two ideas
► a fact or example to support or illustrate an idea
► an alternative interpretation of what you have said before
► a comparison between two places
► a reference to data provided in the question

Please do not feel that you should pad out your answer, repeat something in other words, or write extended lists of names or places. Such filler material gains no marks, and wastes your time when you could be writing something of interest.

The development section must include **examples**. These should:

► Be relevant to the question.
► Be described concisely, but in as much detail as is necessary.
► Be located as precisely as possible. Do not use 'industry in America' if you mean 'industrial areas in the USA'. Do not just refer to 'the USA' if you could be more precise and refer to 'the industrial cities of the northeast seaboard', and if you are referring to those cities you should add '… such as New York and Baltimore', but you do not need to go on and add six other names to this list.
► Include precise details, such as climate figures, or names of the suburbs in a town, or the discharge figures for a stream in flood. You do not need long lists of details, but one or two precise facts are always better than vague generalisations.
► Show a 'sense of place'. You should describe a place so that it would be clear which place you are talking about, even if the name was missed off. Many candidates writing about spontaneous settlements in Bombay write in such general terms that what they describe could be about Sao Paulo. Show that you know what makes Bombay special. Refer to the site on a long, narrowing peninsula; mention the housing near the nineteenth-century textile industrial area; describe the origins of many of the migrants in the neighbouring state of Maharashtra; and so on. Write about a real place, not about generalisations. Examiners get tired of reading about the 'geography of nowhere-in-particular.'

Whenever possible, include **maps** and **diagrams** in your development section. Examiners' reports often comment on how disappointing it was not to see more use of maps and diagrams. Time and again they urge next year's candidates to use them; and time and again the candidates fail to take that advice. So now I urge you again, and here are my reasons:

► Maps and diagrams save time. If you have revised them well you ought to be able to put down a lot of information very quickly on diagrams.
► Maps link information with places, and geography is all about places.
► Maps and diagrams summarise information.
► Diagrams show links, connections and interrelationships clearly and simply, and geography is all about links, connections and interrelationships.

However, to get the most from maps and diagrams remember:

► Always label diagrams, and always give keys to the symbols you use on maps.
► A sketch map will not be perfectly to scale, but the scale should be reasonably accurate. Practise accurate sketch mapping during your revision.

▶ Always refer to your map or diagram in the text. Link it in to the development of your answer, but ...

▶ ... you do not need to repeat information from your diagram in the text. If your map shows where the Cheshire chemical industry found its raw materials and its markets, you need not spend the next paragraph writing this out again.

The **conclusion** is very important. It should not be a repeat of what has already been written. That is a total waste of your time. Of course you may want to draw together points that you have made in different paragraphs of the development; or you may want to summarise a complex argument; or emphasise a key point; but do not just repeat yourself!

The best conclusions refer back to the questions that were posed in the introduction, and answer those questions. Here you may 'assess the extent to which ...'; or you may conclude a discussion, coming down on one side or the other, or even give reasons for remaining undecided; or you may make some reasoned prediction about what might happen in the future, or how a region might develop. Do try to round off your essay and make it feel finished, or concluded.

But what if, despite your careful planning, you run out of time? You could finish off your answer in point form. Try to stick to a structure even then. Stay logical, even if your English style has to be sacrificed. Alternatively, you could refer back to the essay plan. A good, clear plan is often worth several marks, even when the essay has not been written out properly.

Decision-making exercises

Decision-making exercises (or DMEs) combine some of the features of data response questions with features of essays, and even some of the features of coursework. They are actually very like the kind of 'report writing' which many of you will have to do later, on university courses, or in your working life. They are an intense and concentrated form of geographical enquiry.

Some DMEs provide candidates with an information booklet some weeks in advance of the examination, and allow them to study it carefully before they sit the examination. However, the question is only given to the candidates on the day of the examination. Other boards present the information at the start of the examination and do not allow any prior study.

In either case, the process of tackling the exercise is more or less the same. You will have gone through this process of geographical enquiry throughout your school geography career. In Years 8 and 9 pupils do enquiries, but usually in a much simplified and shortened form. Your GCSE coursework, and A-level coursework exercises were probably enquiries that took weeks, or months. Now you have to go through an accelerated enquiry process, in 1½ to 2½ hours. Stay calm! Work logically!

In this type of examination question, even more than any other, you must plan your time efficiently. If you refer back to the idea of planning a journey (on p. 5) it is clear that you have to plan your route just as carefully with a DME – but here you have farther to go, so careful timing is even more vital. It may well be worth making out a time chart. Even if you do not do this during the examination it is worth having one in your mind before the examination starts. It should look something like this:

Time available	Task
4 mins	Identify the task set, by reading the question set and underlining key words and phrases.
30 mins if data not seen in advance	Familiarise yourself with all the data. You will probably want to highlight and note down key references.

10–15% of remaining time	Plan your answer. Make sure you include:
	    relevant techniques
	    references to data

85–90% of remaining time	Write answer. Remember, the structure is usually: present data ⟶ usually fewer marks, so less time. evaluate data ⟶ often most marks, so more time. make a clear decision ⟶ v. important, must not be rushed.

One essential technique to use in a DME is to start thinking about what needs to be done in the last section while you are doing the earlier sections.

> The collection and sorting of information ...
> ... leads naturally to ...
> ... examining several alternatives ...
> ... which leads to ...
> ... making a decision about which alternative to choose.

So you should bear in mind the need to make a decision while you are doing the earlier stages – just as you should refer back to the early stages while you make your final decision.

Finally, remember that this is called a decision-making exercise, and so you have to make a decision. That may seem obvious, but every year far too many candidates are unable to make a clear decision. Quite rightly they have learnt, throughout their A-level geography course, that they must appreciate both sides of a question, and weigh up the advantages and disadvantages, and be aware of other people's points of view, and so on; and it is right to do that in the early parts of the DME. But finally a time comes when you must make up your mind; when sitting on the fence is not enough; when you must **decide**.

If you have been sensible during the early parts of the exercise you will have been working towards the decision, by steadily narrowing down the options. It may be quite easy to rule out some of them. By the end of the exercise, you should have reduced the options to two or three, which have positive support from what you have written in the earlier sections. Then you must make a final decision and **support your decision** with a clear, decisive summary which shows where, in your opinion, the balance of advantages lies.

ASSESSMENT OBJECTIVES IN GEOGRAPHY

All geography syllabuses must award marks for Knowledge (K), Understanding (U) and Skills (S). The proportion of marks for each of these varies across the syllabuses. Each question has marks allocated to either K and U, or S, or both, but all geography examiners agree that it is very difficult to separate out the three different objectives. The mark schemes have to be fairly flexible to allow for this.

However, all candidates are expected to have the following:

► **knowledge of:**
  - geographical terminology
  - the locations and characteristics of places
  - the processes responsible for the development and nature of places
  - geographical ideas, concepts, principles and theories
  - the interaction of people and environments
  - sources of geographical information
  - the processes and techniques of geographical investigation
► **understanding of:**
  - the nature of, and interaction between, physical and human processes
  - the distinctiveness and interdependence of places
  - how physical and human processes bring about changes in places and environments
  - the role of values, attitudes and decision-making processes in geographical issues and the management of resources and environments
  - the potential and limitations of evidence, concepts and theories used by geographers
► **skills**; an ability to:
  - plan and carry out geographical investigations
  - identify, select and collect data from primary sources (including fieldwork) and secondary sources

   – organise, record and present evidence
   – describe, analyse, evaluate and interpret evidence, and draw conclusions
   – evaluate methods and sequences of enquiry, and the conclusions drawn

This description of the assessment objectives really describes the nature of the subject of geography. You should look for opportunities of achieving these objectives in all parts of the examination. Many of the skills may seem to be more appropriate to coursework enquiries, but (with the exception of collection of fieldwork evidence) all of them have a place in the examination as well.

    The examiner asks candidates to show various aspects of their knowledge, understanding and skill by using command words. These words must be clearly understood, and then the commands must be followed precisely. When candidates misinterpret the commands they usually fail to provide what the examiner is looking for – and so they fail to gain marks!

GLOSSARY OF COMMAND WORDS USED IN GEOGRAPHY EXAMINATIONS

Describe Write about what is there.
e.g. 'Describe the distribution of' ...where are they? ... North, South etc .... on the coast .... clustered together ... dispersed ... on the high land
e.g. 'Describe what the diagram shows' ... say what is there on the diagram ... pick out the most important features ... can you see any patterns? ... try not to merely 'lift' information, or repeat it word for word.

Outline Describe briefly.

Give an account Is much the same as 'Describe'. You may be expected to describe changes or developments over a period of time.

Explain Give reasons for what you have observed. This is often linked as 'Describe and explain ....' so having said *what* is there, you go on to say *why* it is there. It is often useful to explain one distribution by referring to another. For instance, explain the distribution of a vegetation community by referring to the distribution of a soil type, and saying how soil affects vegetation.

Suggest This is like 'Explain' but is a bit less certain and more open ended.
e.g. 'Explain why ...' probably means there is a clear reason which can be supported by evidence available. 'Suggest why ...' probably means there are one or more possible reasons, all of which have some evidence in their favour, but none is obviously the only right answer.

State Give a fact, or a reason, or a number of facts or reasons. This usually asks for a short answer, a sentence, or a paragraph at most.

Give This means the same as 'State'.

List State, or give a list of places, processes or whatever. The list may have to be in order, as in a league table, or it may just be random. However, the command 'List' will almost always be followed by some higher level skill, so you may as well structure your list as well as you can while you are doing it. Put the items in order, or arrange them in groups of linked items. It will probably save you time later.

Discuss In a discussion two (or more) sides of an argument should be presented. Each side of the argument should be given in the most logical way possible. So you discuss by saying 'on the one hand this could be the reason for something, but on the other hand that could be the explanation.' A logical argument is one where you start with an idea, and then say 'if that is true, then this will be the result' or 'this happened because of that fact.' You link your points together in a clear sequence.

    Should a discussion reach a conclusion or a decision? Sometimes it will, but not always. At the end of a discussion in an essay it is perfectly acceptable to remain unsure. You may not have enough evidence to reach a conclusion. However, if you do have strong views do not be afraid to express them, as long as you support them with reasons. In a DME you must reach a conclusion – or make a decision – but supported by evidence.

Evaluate Say how much value you give to one or more things. How good is it? What value do you put on it? Here you must reach a conclusion or make a decision.
e.g. 'Evaluate the evidence.' How good is the evidence? Can you rely on it? Is it useful?
e.g. 'Evaluate the map.' Does it show the information clearly? Does it distort evidence?
e.g. 'Evaluate the possible solutions.' Which of the solutions seems to be best? Which will do most good? Which will do least harm? Which best fits the brief?

Analyse Break it down into the parts that make it up. Present the results of the above process. Use the results to discover the general principles underlying the phenomenon you are analysing.
e.g. 'Analyse the information.' Sort out the information into clear sections and present it clearly. Then look for the underlying pattern in the information presented, and possible causes of that pattern.

Compare Say how two things are alike, and how they are not alike. Look for similarities and differences.

Contrast This is similar to 'Compare', but it is really only asking for differences between two places or processes. The command is often linked with another as 'Compare and contrast...' or 'Describe and contrast...'.

Summarise Present the main points in a brief way. Condense the information – cut out all but the most important points. This often means give the ideas but leave out the examples and elaboration.

Consider In an examination this really means much the same as 'Evaluate', but it seems to be less precise.
e.g. 'Consider how modern developments in transport have affected urban structures' means give the information about which developments may have affected structures, and then try to weigh up how influential they have been in causing the changes.

Assess Very similar to 'Evaluate'.

Annotate Put notes on a map or diagram. When you are asked to do this you will *only* gain marks for annotations. The idea is to link facts or ideas with places on the map or diagram. The question is concerned with space and distributions. You will not be given credit for a section of text written below the diagram, unless you make it into an annotation by linking it to a place on the map by using an arrow.

Identify Pick out. Usually this word is used when you have to extract information from data provided in the paper.

Justify Give reasons to support what you say, or to support your choice.

Define Say *exactly* what a word or phrase means. This is usually applied to key ideas from the syllabus. You should have learnt these definitions, and should not have to rely on working them out in the examination.

Comment on This is a rather vague command. It usually means, pick out something from the data given, and say what is interesting about it.

part II
Topic areas, summaries and questions

Ecosystems

In the past geographers studied three quite different topics, called:

▶ climate
▶ vegetation
▶ soils

The division was always rather a difficult one to make, because the three topics were so closely interrelated. Each topic influenced, and was influenced by, each of the other topics, as illustrated by the following little model (Figure 1.1).

So now, several of the boards have a topic called ecosystems, which combines aspects of all three. The precise area covered by the ecosystems topic varies from one board to another. Some limit the study to particular regions; and others say that candidates must study one forest ecosystem and one grassland ecosystem.

There are some boards which still keep climate, soil and vegetation separate, but even these have some part of the syllabus where the different elements are combined and the system has to be considered as a whole. On the other hand, where the syllabus concentrates on the ecosystem approach, there is still a need to study the underlying processes of climate and soil formation separately at some stage. Revision of climate is considered in Chapter 4, but soils and vegetation are considered here, before going on to look at ecosystems.

Figure 1.1

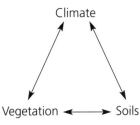

Soils

Describing soils

Soil is a mixture of minerals (broken rock remains), organic material, water and air.

Descriptions of soil usually involve its **texture**, and its **structure**. Texture depends on the proportions of different types of parent material (sand, clay and silt). Structure depends on how the parent material combines with organic material to form crumbs. When there is a good proportion of organic material in the soil it is said to have a crumby structure. Such a soil is easy to work and holds plenty of water and nutrients for plant growth. Soils with a low proportion of organic materials have a blocky or platy structure, and are less suitable for farming.

Soil development

Sometimes it is possible to study the formation of soils on newly formed, virgin land surfaces, such as newly erupted volcanic rock, or rock exposed by a landslide. Soil formation involves the addition of organic matter to the minerals from the rock; and the vertical movement of minerals in the soil, mainly caused by movement of water. Figure 1.2 overleaf shows the movement of organic matter and water through the soil. When water removes soluble salts from a horizon this is called **eluviation**. When the salts are deposited elsewhere in the soil profile it is called **illuviation**. If insoluble particles are moved downwards the process is called **translocation**.

The rate of movement of water through the soil, and its direction (up or down) depends on the balance between precipitation (P) and potential evapo-transpiration (Pet). This is summarised as the **P:Pet ratio**. If P>Pet, net movement of dissolved minerals is downwards: when P<Pet, net movement is upwards.

Figure 1.2

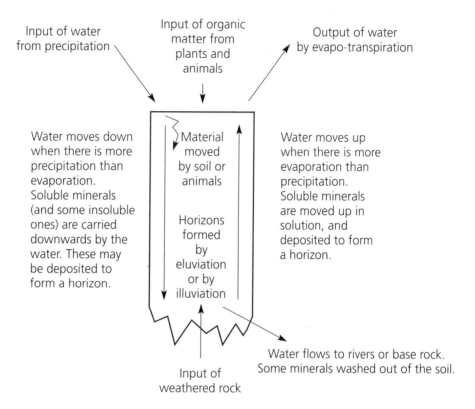

Input of water from precipitation

Input of organic matter from plants and animals

Output of water by evapo-transpiration

Water moves down when there is more precipitation than evaporation. Soluble minerals (and some insoluble ones) are carried downwards by the water. These may be deposited to form a horizon.

Material moved by soil or animals

Horizons formed by eluviation or by illuviation

Water moves up when there is more evaporation than precipitation. Soluble minerals are moved up in solution, and deposited to form a horizon.

Water flows to rivers or base rock. Some minerals washed out of the soil.

Input of weathered rock

The main factors affecting soil formation can be summarised under the rather ugly mnemonic CLORPT.

CL	= Climate	which affects	– weathering of parent material – growth and decay of vegetation – amount and activity rate of soil organisms – eluviation and illuviation
O	= Organisms	including	– plants which produce biomass – animals consuming and decomposing biomass – the often crucial human influence

R	= Relief	which affects	– drainage
			– micro-climatic conditions
			– mass wasting causing:
			    loss of soil from some areas
			    accumulation of soil in others

P	= Parent material	which affects	– texture (gritty, sandy, clayey, silty) – depth, due to speed of weathering – pH value – acid or alkaline
T	= Time		– at first a soil is mainly parent material – eventually climate becomes the dominant influence (zonal soils) – but in between these stages soils form which are not in equilibrium with climate (azonal soils)

Zonal soils are linked with particular climate regions or **biomes**. Each of these has a characteristic combination of horizons, or **profile**, and you should learn the main profiles of soils on your syllabus. Most soils have A, B and C horizons:

▶ **A horizons** or **topsoil** usually contain organic material, but often they have lost material through eluviation.
▶ **B horizons** or **subsoil** are usually layers where minerals accumulate through illuviation.
▶ **C horizons** are mainly weathered bedrock, or regolith, which has not been altered much by soil forming processes.

Three processes that you should be particularly aware of are:-

▶ **Leaching** Where low to moderate rainfall and a high P:Pet means that the more soluble minerals (N, Ca, Mg, Na and K) are removed from the A horizon and deposited in the B horizon.

▶ **Podsolisation** Where the rainfall is high and the P:Pet is high. Acid humus forms, and water passing through this becomes acid. The more soluble minerals listed above are dissolved, and washed out of the soil. Even iron and aluminium salts are dissolved from the A horizon, and may be precipitated in the B horizon. This illuviation may form a hard, impermeable layer.

▶ **Gleying** Where impeded drainage means that water cannot drain away. The soil becomes waterlogged. This leads to chemical reduction of the iron salts in the soil, forming grey, ferrous compounds. The grey soil may be mottled with red patches where bubbles of air reduce the ferrous salts to ferric salts.

Azonal soils form where some factor impedes the full development of the zonal soil. It may be impeded drainage, leading to the formation of a gley; or rapid mass wasting may remove so much material that a skeletal soil is produced.

Vegetation

Vegetation structures
The vegetation of an area is the overall grouping and arrangement of plants. When there is just a single species present in an area it is called a **plant society**. This is most unusual, though, and when there are two or more species in the area it is called a **plant community**. When a community has a main, **dominant species** it is called a **plant association**. A community usually extends across an area which has consistent soil and climate characteristics. However, similar communities with slight variations between regions are found stretching across whole world climate regions (such as the equatorial region). Such communities, linked with particular climate and soil types, are called **biomes** (e.g. the equatorial rainforest biome).

A number of environmental factors usually interact to produce vegetation. They are summarised in the diagram below which shows that a vegetation community reacts to outside, and internal influences. Over time the community develops, eventually reaching a state of **dynamic equilibrium**.

Figure 1.3

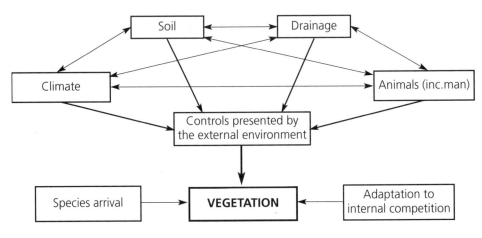

Operating through time

Vegetation successions
Newly exposed ground will soon be invaded by plants. These early, colonising plants are called **pioneer species**. By growing there they cause changes to the surface, developing the soil (stabilising it and adding organic matter), altering the micro-climate, and allowing other species to replace them. This is known as a **plant** or **vegetation succession**. Gradually other species move in as part of the succession until a **climatic climax community** is established, which is more or less in equilibrium. The whole process is called a **sere**. Each group of species in the **seral progression** is called a **seral stage**.

As a succession proceeds there is usually an increase in:

- ▶ the number of species in the community (at least in the early seral stages)
- ▶ the complexity of the community (often shown in increased layering)
- ▶ the biomass (or weight of organic material)
- ▶ the productivity of the community

When a sere happens on a newly exposed rock surface it is called a **primary succession**. If it takes place on a surface that was formerly covered with vegetation which was cleared by cutting, burning, etc., it is called a **secondary succession**. Sometimes a succession may be interrupted by a local factor which stops the progression towards the climatic climax. This produces an **interrupted succession**. Examples of interrupting factors, and the communities produced, are listed below:

Interrupting factor	Type of community
Local soil	edaphic or soil climax
Relief	topoclimax
Impeded drainage	hydroclimax
Animal interference	biotic climax
Human interference	plagioclimax

Human interference often destroys an established climax community. The alteration of the community is called a **retrogressive succession**.

Four kinds of primary succession are generally recognised as occurring on different types of new surface:

- ▶ lithoseres on bare rock or rock slides
- ▶ psammoseres on sand, especially sand dunes
- ▶ hydroseres in ponds or lakes
- ▶ haloseres on salt marshes

Each has distinct characteristics and sequences. It is important that you know (or at least recognise) the ones needed specifically for your syllabus. It is especially useful if you can remember examples that you have studied in the field, or as detailed case studies.

The structure of ecosystems

An ecosystem consists of a group of living organisms (animals and plants) and their physical and chemical environment (particularly the atmosphere and the soil). Within the ecosystem there is biomass (weight of living matter), usually given as tonnes/hectare or kg/m^2. This is present below ground (as roots, micro-organisms, etc.) as well as above; and there is also dead organic matter (DOM), made up of surface litter and humus. Ecosystems are sustained by the **flow of energy** through them, and by the **nutrient cycles**. The original source of all energy is from the sun.

Figure 1.4

The energy from the sun is fixed by green plants, or primary producers, (also known as autotrophs) through the process of photosynthesis. Some of the plant tissue produced is then eaten by animal herbivores, known as consumers. They may be consumed by carnivores and the lower carnivores may, in turn, be eaten by top carnivores. However, much plant tissue is not consumed, but (along with animal waste) enters the soil, where it forms humus. In the soil, plant and animal waste can be used, and broken down by micro-organisms called decomposers.

A simple, linear relationship between the different producers and consumers in an ecosystem forms a **food chain**; each stage in the chain, where food energy is exchanged, is called a **trophic level**, e.g.:

Figure 1.5

Grass   Cattle   Humans

T_1 ⟶ T_2 ⟶ T_3 (T=Trophic level)

Plant   Herbivore   Carnivore

Simple, linear, food chains can exist in nature, but normally ecosystems are more complex and **food webs** exist, with many interactions. For instance, maize could be grown to feed both cattle and humans; the humans may consume both maize (as herbivores) and the cattle (as carnivores). However, in any ecosystem there will always be more individuals in the lower trophic levels than in the higher ones. This is because much of the energy in the lower levels is lost as it moves through the system. The diagram below represents the number of individuals in the trophic levels of an ecosystem:

| Lions |
| Zebras |
| Grass plants |

In some natural ecosystems all the nutrients produced are recycled, so that the system is in perfect balance. However, an ecosystem can be upset, so that nutrients are removed and not replaced. Human intervention is the most common cause of such imbalance. At a simple level a shortage of firewood may lead to burning of animal dung and the loss of the dung as a soil nutrient. On a much larger scale the world trade in food may lead to the rapid depletion of the soil in some areas that are farmed on an exploitative basis.

★ REVISION ACTIVITIES

Figure 1.6

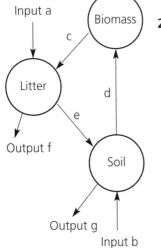

Input a

Biomass

c

Litter

d

e

Output f

Soil

Output g

Input b

1 The diagram to the left (Figure 1.6) is a model of the mineral nutrient cycle. Copy it and label the nutrient transfers.

2 The three diagrams below represent rainforest, steppe and coniferous forest ecosystems. The size of the compartments and flows is proportional to the flows and stores. Which is which? Write notes to explain the differences.

Figure 1.7

(a)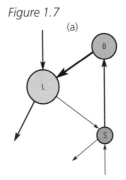

(b)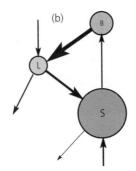

(c)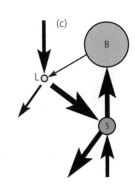

? PRACTICE QUESTIONS

Question 1

Study Figure 1.8 overleaf, which shows a system of trophic levels (a food chain).

(a) Identify: the input X
                the flow Y
                the flow Z   [3 × 1 line] (3)

Figure 1.8

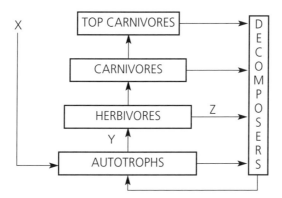

(b) Explain:
   (i) The way that energy conversion occurs between X and the autotrophs.   [4 lines] (3)
   (ii) Why each higher level in the system reduces in size/number.   [4 lines] (4)

(c) (i) Suggest why the system shown in Figure 1.8 might be considered
       as unstable.   [3 lines] (2)
   (ii) State two results of this instability.   [4 lines] (2)

(d) With reference to your chosen case study area, describe with the aid of an annotated
    diagram, how human activity can alter the trophic level in an ecosystem.
                                           [6 lines + space for diagram] (6)

                                                                        (*London*)

Question 2

Describe and explain the natural processes of soil formation that characteristically
occur in temperate environments, and discuss how human interference might modify
these processes.   (25)

                                                                        (*Oxford*)

Question 3

Explain, with reference to *two* contrasting seres, why ecosystems are described as
dynamic systems.   (20)

                                                                        (*WJEC*)

Question 4

You have been asked to study the soils of a 100 km^2 area within a drainage basin.

(a) Describe the methods you would use to sample the soils of the area.   (7)
(b) At each sample point characteristics that might be studied include soil texture, soil
    structure, soil acidity and organic content.
   (i) Choose any *two* soil characteristics and describe how you would measure and
       analyse *each* of them.   (4)
   (ii) Describe how you would expect soil characteristics to vary within the drainage
       basin, e.g. on a transect from the watershed to the valley floor.   (7)
(c) Discuss the ways in which human activity may influence soil characteristics.
                                                                                   (7)

                                                                        (*NEAB*)

2 River systems

A river system is one part of the much bigger system, the **water cycle**. Water that falls as precipitation is **transferred** through the river system back to the sea. Some water may be lost from the system (particularly through evapo-transpiration), and water may be **stored** for short or long periods, but the fundamental movement in the system is the transfer of water, under the force of gravity, down to the sea.

The system consists of:

▶ precipitation: rain, snow, hail, sleet, dew, mist, frost
▶ transfers: rivers, overland flow, throughflow, etc.
      managed transfers, (e.g. pipes, canals, sewers, etc.)
      take-up by vegetation
▶ storages: ice caps, snow fields, puddles, lakes, etc.
      soil water, ground water, etc.
      managed storage (e.g. reservoirs, tanks, etc.)
▶ outputs: to the sea
      evaporation, transpiration

A **river basin** is the whole area drained by a river and its tributaries. The basin is surrounded by a **watershed** or area of high land. The watershed divides the area draining to one river system from the area draining to another system. River basins are often likened to leaves. The veins on a leaf look rather like the streams in a river basin, all feeding into the main river. However, even the area between the streams is part of the drainage system. Water drains over the surface, or through soil and rock, until it reaches the river.

The **fluvial system** is rather more complicated than the drainage basin. Whilst the drainage system just considers flows of water, the fluvial system takes **sediment** into consideration too. Models of the fluvial system show three parts:

▶ Zone 1 is the area close to the watershed, where sediment is produced by weathering and erosion
▶ Zone 2 is the transportation zone; sediment is moved through this area by the river
▶ Zone 3 is the deposition zone

In real life, although one process is usually dominant in each area, there is some overlap between the processes.

The **water balance** in a river basin is the balance between inputs of water from precipitation, and losses by evapo-transpiration. The water balance obviously has a big effect on the flow of streams in the basin, although it is not the only factor. The actual flow of streams can be shown by **river hydrographs**. These show how the river's flow, (usually measured in cubic metres/second, or cumecs), varies over a period of time. Storm hydrographs, which show the flow before, during and after a storm event, are particularly important for geographers. Figure 2.1 shows an idealised storm hydrograph.

Storm hydrographs show the speed with which a flood follows a storm, and the height or intensity of the flood. A short **lag time** and a high **peak flow** are produced when water transfers to the river take place quickly. The following are some of the conditions that can contribute to a rapid flood:

▶ a large proportion of rainfall as run-off: a low proportion as throughflow
▶ little vegetation:

- to intercept rainfall
- to slow down run-off
- to take up water from the soil
- often as a result of land management practices
▶ reduced infiltration due to:
  - a hard baked soil surface
  - a thin soil layer
  - soil already saturated by previous rainfall
  - impermeable bedrock
  - buildings making surfaces impermeable
▶ fast run-off due to:
  - sudden, heavy rainfall
  - steep slopes
  - human interference with drainage of the soil, etc.

Figure 2.1
Idealised storm
hydrograph

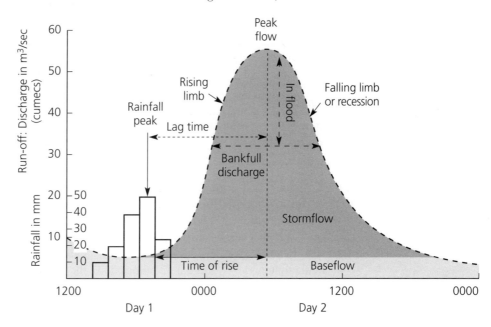

Much of geographers' work on river systems is concerned with the characteristics of **river flow** and the way in which this influences **channel shape**. The nature of the river and channel generally change as the river moves downstream. It is useful to summarise some of the differences between the upper course and the lower course. The table on p. 20 provides a useful checklist, but you need to be able to fill this out with detail before you enter the examination. A thorough revision programme will ensure that all (or almost all) of the points on the table are understood in as much depth as possible. Then detailed case study material to illustrate the ideas will have to be learnt. The table is a structure; you must add finishing touches of your own.

The erosion and transportation processes listed in the table cause gradual lowering of river beds. They also cause valley sides to be lowered, as they remove weathered debris from the banks. If nothing interrupted the lowering of the bed and sides the level of the valley would eventually be reduced to **base level**. This is the height at which the water in the river system finally loses all its potential and kinetic energy. Base level is at sea level in most river systems, and rivers cannot erode below this height. The diagram below shows how the river long profile would gradually approach base level throughout its length – if nothing else affected the valley.

Figure 2.2

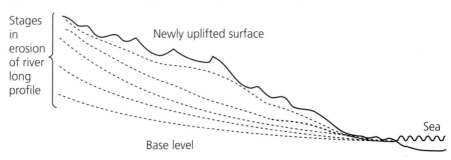

		Upper course	*Lower course*
The Valley	Gradient	Steep	Gentle
	Width	Narrow	Broader
	Depth	Shallow	Deep
	Cross section	Narrow V-shape	Wide flood plain with bluffs
The Channel	Velocity	Irregular, averages slow	Regular, faster
	Discharge	Low volume	Large volume
	Turbulence	Turbulent	Smooth flow
	Sinuosity	Fairly straight	Meandering
	Cross section	Rough, irregular	Smooth, rounded
Erosion	Capacity	Low	High
	Competence	High, in flood periods	Low
		Sand, boulders	Clay, silt
	Main processes	Erosion: corrasion, attrition, hydraulic	Mainly transportation: suspension, solution
		Some transportation: mainly bedload	
Deposition	Type	Unsorted bedload deposited after floods	Sorted suspension load deposited on bed in low flow periods, or on flood plain after floods
	Shape	Angular	Rounded
Selected features		Potholes, waterfalls, interlocking spurs, gorges, etc.	Flood plain, meanders, point bars, ox bows, deltas, etc.

At any time during this progression the river system may reach a state of of balance or **dynamic equilibrium**. Any river has a certain amount of energy, provided by the force of the water flowing downhill under gravity. When the system is in balance the available energy is used up in moving the sediment that is brought down to the river by mass wasting. This means that the channel shape is kept fairly constant, input being matched by output at all points along the river's course. The valley, as a whole, will be slowly lowered towards base level.

The system may be upset by the sudden input of a larger than usual amount of sediment. This could be caused by a landslide, a volcanic eruption, or by human activity, such as dumping of mining waste in the valley. Such an event provides a disturbance to the system. It will be followed by a period of recovery. The river has to use energy to remove the extra sediment, and this is likely to lead to changes downstream as the extra sediment is removed, but then the progress towards base level is re-established.

However, land is rarely stable long enough for base level to be reached along much of any river's course. The relationship between land and sea level is constantly changing. **Tectonic** processes can cause the level of the land to rise and fall. **Eustatic** processes cause sea level to rise and fall. Most of the world's coastal areas have been affected, in the last million years, by eustatic changes linked with the formation and melting of the ice caps during the ice ages. Each change in sea level causes a change in base level, which results in changes throughout the river system.

When base level falls the river is **rejuvenated** and gains increased energy for downcutting. This produces a characteristic set of erosional features such as river terraces, nick points, incised meanders, etc. The renewed downcutting may also lead to **river capture**. If sea level rises relative to the land, the river loses energy, and deposition occurs in the lower stages of the river and in the estuary.

The **drainage pattern** of a river system describes the overall plan of the river and its tributaries. Drainage patterns were originally descriptive and rather subjective. Some of the more common patterns are shown in Figure 2.3.

Various systems of **stream numbering** were worked out to try to make description more scientific and objective. One such pattern is shown in Figure 2.3. From this it is possible to work out the number of stream segments of each order, bifurcation ratios, drainage densities and so on. The ideas are useful for hydrologists,

Figure 2.3
Selected stream
patterns and a stream
numbering system

Dendritic

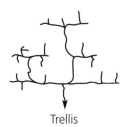

Trellis

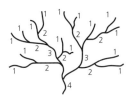

Parallel

Stream numbering

but the actual measurements are rather complex, and beyond the scope of most A-level geography syllabuses.

Some years ago there used to be a clear division between physical geography and human geography. The study of rivers was clearly part of the physical side of the subject. Nowadays, fortunately, the syllabuses give full emphasis to the links between the two parts of the subject. You must be aware of the way that rivers influence, and are influenced by, human activities. Sometimes the influences are direct and deliberate; at other times they are indirect and accidental.

Among the many ways that rivers affect human activities are:

▶ providing water for agriculture and settlement
▶ threatening those same fields and settlements with floods and/or erosion
▶ providing transport routes for boats
▶ cutting valleys to provide routeways for roads, railways, etc.
▶ presenting barriers to transport which has to cross rivers
▶ attracting settlement to the points where routes can cross the rivers
▶ providing opportunities for leisure
▶ providing opportunities for fishing
▶ acting as sewers and places for waste disposal
▶ polluting areas downstream, below waste disposal points

And humans affect the rivers by:

▶ draining land and speeding run-off
▶ taking water from the rivers for agriculture, industry and domestic use
▶ attempting to control erosion
▶ canalising rivers to allow inland water transport
▶ dredging estuaries to ease the route for sea-going transport
▶ damming the head waters for storage reservoirs
▶ heating streams by adding power station cooling water

First you must learn your detailed examples showing how physical processes operate, through time, to produce landforms in valleys; then it is time to understand and learn examples showing how the processes and features interact with people.

REVISION ACTIVITIES

1 The diagram shows a storm hydrograph.

Figure 2.4

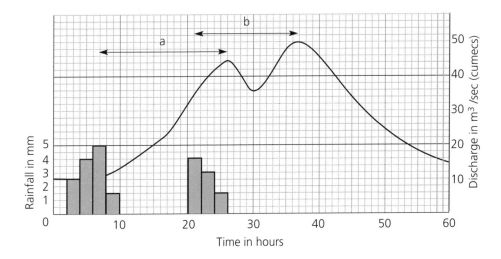

(a) What is the total rainfall during the first storm?
(b) What is lag time a?
(c) What is the maximum flow during the first flood?
(d) What is the total rainfall during the second storm?

(e) What is lag time b?

(f) What is the maximum flow during the second flood?

(g) Suggest reasons for the differences between the two floods.

2   In each of the cases below, imagine that there are two identical river basins, with only one aspect different. In each case, say which river is more likely to flood, and explain why.

	Basin A	**Basin B**
Case 1	underlain by permeable rock.	underlain by impermeable rock.
Case 2	thick soils.	thin soils.
Case 3	steep slopes.	gentle slopes.
Case 4	recently deforested.	mature forest.
Case 5	dendritic drainage pattern.	parallel drainage pattern.
Case 6	car park built on grassland.	no car park – just grassland.
Case 7	snowfields face south – so melt suddenly in spring.	snowfields face north – so melt slowly in spring.
Case 8	farmers plough around contours.	farmers plough straight up and down slopes.
Case 9	moorland peat ploughed to allow planting of forest.	moorland peat undisturbed.
Case 10	rapid erosion on upper slopes.	slow erosion on upper slopes.

? PRACTICE QUESTIONS

Question 1
With reference to your located case study, discuss the interrelationships between factors influencing river discharge.   (20)

(London)

Question 2
Either A

(a) With respect to a named drainage basin, and with the aid of a diagram or diagrams, describe and explain the seasonal variations in river flow.   (13)

(b) Describe and explain how you would investigate downstream variations in discharge and velocity of river flow.   (12)

Or B

(a) With reference to a named drainage basin, explain how it might be regarded as an integrated physical system.   (13)

(b) Explain how you would investigate the relationships between slope form, soil and vegetation characteristics in a chosen part of that drainage basin.   (12)

(Oxford)

Question 3
(a) Describe the hydrological cycle in detail.   (12)

(b) Explain how human actions may alter the amount of water held in the groundwater zone.   (13)

(AEB)

3 *Plate tectonics*

To revise this topic you must understand one big idea. You must appreciate that the earth is not fixed and rigid, but is continuously moving and changing. This is explained by the theory of plate tectonics. Then, you need to learn how this movement causes certain events to happen at the plate margin. Once you appreciate how plate margins can be classified, it becomes reasonably easy to learn what happens at the different types of margin. Then you must combine the understanding of the physical processes with an appreciation of their effects, both positive and negative, short term and long term, on people and their ways of life.

Plate tectonic theory

The most widely accepted estimates suggest that the earth was formed 4 600 million years ago. It has recently become clear that the earth's major surface features, the oceans and continents, have moved back and forth across the surface; they have only recently reached their present positions; and they are likely to continue to move in future.

The theory of plate tectonics was only developed in the mid-1960s, and very quickly became almost universally accepted by scientists. The full details of the theory probably lie in the field of geology, but a basic understanding is necessary for geographers. Because of the nature of the subject, and the importance of people-environment links in geography, we need to understand enough geology to know what is happening to the earth, and why.

The earth's structure

The earth is formed from a number of concentric shells of matter (Figure 3.1). In the centre is the **core**, formed from very dense, very hot (about 5 500 °C) rocks. Parts of the core are probably solid, parts are semi-molten. The decay of radioactive material in the core probably generates the heat which powers the movements of the plates.

Around the core is the **mantle**. This section is also hot (up to 5 000 °C) and dense, but less so than the core. The inner parts of the mantle are semi-molten, and move slowly under the influence of convection currents, caused when the mantle rocks are heated by the core rocks. The top part of the mantle is rigid.

The **crust** is the outer shell. The rocks forming the crust are lighter and cooler than the deeper rocks. They are solid. Together with the rigid rocks of the upper mantle they form the plates. There are two types of crustal plate, the **oceanic plates** and the **continental plates**.

▶ Ocean crust is formed of denser rock. It is **basaltic** consisting mainly of silica (Si) and magnesium (Mg), and is sometimes referred to as Sima.
▶ The continental crust is formed of older and less dense rocks. The rocks are **granitic**, formed largely of silica and aluminium (Al), and are referred to as Sial.

Plate movement

Various theories of continental drift have been put forward to try to account for facts like the apparent 'jigsaw-like' fit of some of the continents. These were never taken very seriously by the scientific community until the 1960s. Plate tectonic theory developed when evidence was found showing that the floor of the Atlantic was spreading. Study of volcanic rocks and sediments from the ocean bed showed that new

Figure 3.1
The internal structure
of the earth

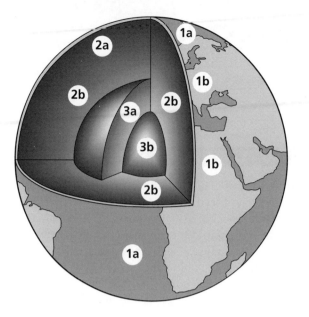

1 **Crust**
 (a) Oceanic crust (sima) mainly
  of basalt, averaging 6-10km
  in thickness. At its deepest it
  has a temperature of 1200°C.
 (b) Continental crust (sial) up to
  70km thick. The crust is
  separated from the mantle
  by the Moho discontinuity.

2 **Mantle** Composed mainly of
 silicate rocks, rich in iron and
 magnesium. Rigid top layer
 (2a), the rest is in a semi-molten
 state **(2b)**. The mantle extends
 to a depth of 2900km where
 temperatures may reach 5000°C.
 These high temperatures
 generate convection currents.

3 **Core** Iron and nickel. Outer
 core **(3a)** is semi-molten. Inner
 core **(3b)** is solid. The
 temperature at the centre of
 the earth (6371km below the
 surface) is about 5500°C.

rock was forming in the mid-Atlantic, and that the ocean was getting wider by approximately 5 cm each year. If new crust was being formed here, and the earth was not getting bigger, then crust must be being destroyed somewhere else; and something must be moving the crust around.

It was then realised that:

▶ The crust formed separate 'blocks' or 'plates'.
▶ These were moved across the surface by convection currents in the mantle.
▶ When two plates met at a **plate margin** a number of possible processes and landforms could be seen. The nature of the process depended on the directions of relative movement at the margin, and the nature of the crust on either side.

Types of plate margin

Constructive margins
These are called 'constructive' because new plate is constructed as magma from the mantle rises to fill gaps on the sea floor. Plates are pulled apart by the action of convection currents in the mantle. These slow moving currents drag on the bottom of the plates, tearing them apart. At first a **rift valley** may be formed on the sea bed when sections of the crust slip down between parallel faults. Soon, however, the release of pressure on the mantle, caused by the rifting, allows the semi-molten rock to become more liquid, and magma rushes to the surface. This forms submarine volcanoes on either side of the rift. These may grow big enough to emerge above the surface, as has happened in Iceland, which lies on a very active part of the mid-Atlantic ridge, and the volcanic zone runs right across the centre of the island. Its recent volcanic history includes three spectacular eruptions:

▶ In 1962 a new island, called Surtsey, was formed off the southwest coast. It is formed of ash, with a hard covering of lava on top, and rises 130 m above sea level.
▶ In 1973 the older, inhabited island of Heimaey, lying between Surtsey and Iceland, had to be evacuated whilst their volcano underwent a six-month long eruption.
▶ Then, in 1996, there was an eruption on the mainland, beneath a glacier. The erupted material did not cause any damage, but the heat melted ice producing millions of gallons of water. This water could not be seen but found ways under the ice; it eventually burst out, flooding to the sea, and carrying huge amounts of unconsolidated glacial deposits. The flood did not affect inhabited areas, but it destroyed major bridges on the main east-west road across the south of the island.

It may be that a new constructive margin is forming in East Africa. The Great Rift Valley, with volcanoes along either side, seems to be forming where the continent is splitting apart. The Red Sea is probably part of the same process. East Africa and Arabia may eventually become detached from Africa, and a new ocean may form.

Boundary type	Type of movement	Landform changes	Example
Constructive.	Two plates move apart.	Volcanoes form mid-ocean ridge.	Mid-Atlantic ridge.
Destructive.	Two plates move together. Ocean crust moves towards continental crust.	Subduction zone. Heavier, ocean crust sinks and is destroyed.	Japanese island arc. W coast of S America.
Destructive also called Collision.	Two plates move together. Two continental crusts collide.	Neither plate sinks. Sediments pushed up to form fold mountains.	Himalayas. Alps.
Conservative, Slip or Passive.	Two plates move side by side.	Crust is neither created, nor destroyed.	San Andreas fault in California.

Destructive margins

These get their name because of the destruction of crust which happens there. The process also causes great destruction of life and property, but that is not the reason for the name!

When an ocean plate meets a continental plate, or when two ocean plates meet, the heavier plate sinks into the mantle. This forms a **subduction zone**. On the sea bed it forms a deep, steep-sided **ocean trench**. As the crust plunges deeper into the mantle, friction with surrounding rocks generates heat which melts the plate material. This molten rock is lighter than the surrounding mantle, and so pressure forces it back towards the surface. Here it can either form:

▶ **Batholiths**, which are huge masses of granite rock which cool below the surface (**intrusive** volcanic rock).
▶ Or **volcanoes** where the magma erupts at the surface and then cools and solidifies (**extrusive** volcanic rock).

During the subduction of the plate, the friction between the sinking plate and the surface plate can hold up the movement. This leads to a gradual build up of pressure as the convection current continues to drag the plate down. Sooner or later the pressure gets so great that the plate moves, with a sudden, catastrophic jerk. This build up and sudden release of pressure causes earthquakes.

At an ocean–ocean destructive margin, such as is found beneath the Philippines, the West Indies and Japan, the subduction process produces a volcanic **island arc**.

At an ocean–continent destructive margin, such as the west coast of South America, the collision also forms a range of fold mountains, formed from sediments off the former sea bed. Fold mountain ranges like the Andes normally have giant batholiths somewhere below the surface, and their surfaces are dotted with lines of volcanoes.

All destructive margins are also subject to earthquakes.

Collision margins

When two plates meet, and both are carrying continental crust, neither can sink beneath the other. The edges of both continents usually have huge deposits of sediments, formed when material eroded from the land is deposited in the bordering seas. When the continents collide these sediments are subjected to enormous pressures, which force them up to form mountain ranges. Earthquakes are common in such regions; volcanoes are less common, but may occur when a sudden release of pressure allows rock in the mantle to become molten and flow towards the surface.

The Himalayas and the Alps were formed at collision margins, where the Indo-Australian and African plates collided with the Eurasian plate.

Conservative margins

At these margins no crustal rock is being created, and none is being destroyed. Two plates slip past each other moving parallel, or almost parallel, with each other. There is no volcanic activity at such margins, but earthquakes are common. As at destructive margins, friction between the two masses of rock can stop the plates moving; but pressure continues to build up until eventually it is released in a sudden catastrophic movement.

California is the best known location of a conservative margin. There are many fault lines associated with the margin, and the San Andreas fault is the best known of these.

What lies away from the plate margins?

On a global scale it is quite evident that interesting processes take place on, or close to, the margins of the plates. At their centres the plates are rigid and little or no activity takes place there; although East Africa is an exception, dealt with above.

There are three main types of continental land area:

1 **Cratons** or **shield areas**, which are formed of old, hard rocks.
   - these are the remnants of fold or volcanic mountain areas which may have been through many periods of uplift followed by erosion
   - structures are often complex, and only the hardest rocks remain
   - the surfaces have often been worn almost flat, although they may be undulating with a mass of lakes and low hills
   - the hard rock usually leaves thin soils which are of little use for agriculture
   - they may contain important mineral deposits, which can attract settlement
   - the Canadian (Laurentian) Shield, the Scandinavian Shield and the Brazilian Shield are all cratons
2 **Lava plateaus** formed when lava escapes from a fissure eruption, forming basalt rocks.
   - repeated flows form plateaus which are thousands of metres deep
   - they sometimes form when plates split apart
   - when these rocks weather they can form fertile soils
   - Greenland was formed from a lava flow, formed when Eurasia and North America split apart; the Deccan plateau in India, and the plateaus to the west of the Rockies in North America are other examples
3 **Major river basins**, which form in depressions, towards the edges of the continents.
   - they contain sediments, eroded from shield areas and neighbouring fold mountain ranges
   - these sediments are often very deep, because the crust beneath slowly sinks into the mantle under their weight
   - they are attractive areas for agriculture and settlement
   - some sedimentary basins contain coal and oil deposits
   - examples include the Mississippi, the Yangtse and the Amazon basins

How events at plate margins influence people

All the above describes events and their causes. If events threaten people they are seen as hazards. Sometimes events occur which have such a devastating effect on people that they cause disasters. Many earthquakes and volcanic eruptions occur without being hazards. Either no one lives near enough to be threatened, or the event can be managed so that people are safe. Only very few tectonic events become disasters.

Unfortunately, as the world's population increases, more and more people are being forced to settle in areas which present serious potential hazards. Moreover, the people pushed into these areas are often the poor, who are least able to prepare for and cope with the hazards. Some people think that the frequency of disasters caused by tectonic events is increasing. If this is so it is probably not because the events are becoming more frequent; rather it is because more people are having to take more risks.

Whenever you have to write about the way that tectonic events affect people you need to structure your answer to include some, or all, of the following points:

1 **What do people expect?** How good is their knowledge of the area and its risks? Can they predict the risk with any certainty?
2 **How do they perceive the balance of risk and opportunity?** Areas with risk often also present opportunities. The fertile soil from previous volcanic eruptions is a prime example. If people have few opportunities they may have to take bigger risks in order to survive.
3 **Can people plan to minimise the danger?** Usually the rich can afford to plan well, and the poor cannot. The Kobe earthquake showed that even rich countries, where planning is supposed to be thorough, can still suffer disasters.
4 **What are the immediate and the long-term effects of the event?** These include:
   - death and injury during the event

– lack of shelter, cold, hunger, starvation, spread of disease, loss of livelihood etc. in the days and months after the event

– cost of rebuilding, grieving, anxiety, increased insecurity, etc. in the months and years after the event

– possible improved life style and adaptation to the hazard in the long-term future

The outline of the ideas above must be adapted and developed to fit the examples you have studied, but the outline provides a structure around which you can organise your knowledge. It provides an outline essay plan for an examination answer.

REVISION ACTIVITIES

Learn to draw these diagrams, and complete the labels to show:

1 constructive margin
2 destructive margin
3 conservative margin

Figure 3.2

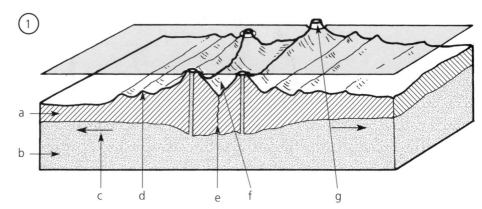

Figure 3.3

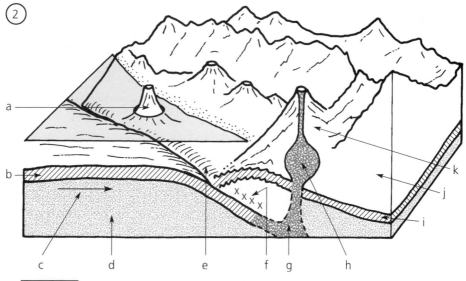

Figure 3.4

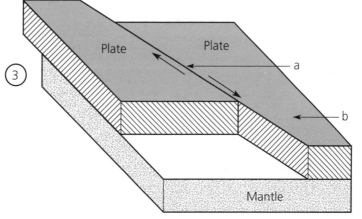

PRACTICE QUESTIONS

Question 1

Study Figure 3.5 which shows part of the world system of tectonic plates and plate margins.

Figure 3.5

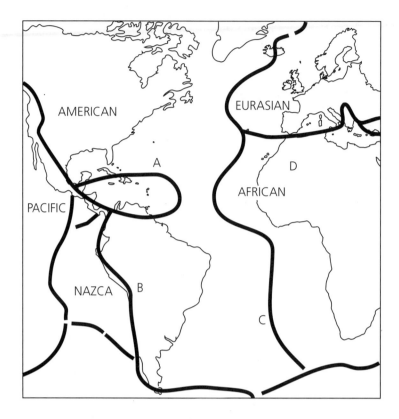

(a) Using the terms 'constructive', 'destructive' and 'conservative' describe *each* of the margins labelled A to D. [4 lines] (2)

(b) Suggest *three* pieces of evidence supporting the view that at one time South America and Africa were joined. [6 lines] (6)

(c) With the aid of an annotated diagram, describe the processes taking place along a destructive plate boundary. [6 lines + space] (6)

(d) Suggest the advantages available to the many people who live on an active plate margin. [6 lines] (6)

(*London*)

Question 2

(a) Assess the evidence which indicates that South America was once joined with Africa. (9)

(b) How does the theory of plate tectonics help us to understand:
(i) mountain building (8)
(ii) earthquake distribution? (8)

(*Oxford*)

Question 3

The map in Figure 3.6 shows the major tectonic plates and their boundaries in the western hemisphere. Part of the key has been left blank.

Figure 3.6

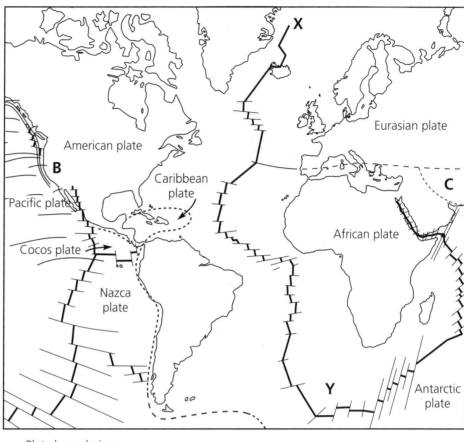

Plate boundaries

——————   ....

══════   ....

----------   ....

- - - - -   Uncertain or inactive

(a) (i)   Complete the key to the map.   (3)
    (ii) Indicate clearly and mark with the letter *A* the location of an island arc on the map. (1)

(b) In the space below, draw a labelled diagram to illustrate the processes occurring along
     the plate margin *X–Y*.   [Space for diagram] (5)

(c) (i)   With the aid of examples, explain *two* benefits which people have gained from the
          presence of plate margins.   [5 lines] (4)
    (ii) Why might the earthquake hazard at *B* be seen differently by the local population
          compared with the same hazard at *C*?   [3 lines] (2)

                                                                                 (AEB)

Question 4

With reference to any *one* type of natural hazard associated with active landscapes,
explain its possible causes and its likely effects on the landscape and on people.   (20)

                                                                                 (WJEC)

4 Meteorology and climate

The study of climate and weather, or meteorology, are obviously closely linked.

▶ **Weather** describes atmospheric conditions at a place, at a certain point in time.
▶ **Meteorology** is the study of the atmospheric processes that produce weather.
▶ **Climate** describes the average conditions at that same place over a much longer period. Usually climate figures are averages taken over at least 35 years.

In all these areas the elements that must be studied are temperature, humidity, precipitation, air pressure and movement of the air. Climate is particularly concerned with the seasonal variations in each of these elements.

The heat budget

The sun is the source of the energy which powers the meteorological system. Energy arrives at the edge of the earth's atmosphere in the form of **short wave radiation** or **insolation**. Some of the energy is **absorbed** by the atmosphere, especially by ozone, carbon dioxide, water vapour and dust. More energy is **reflected** back by the clouds and by the earth's surface. Note that light coloured surfaces, such as ice and sand, reflect more energy (they have a higher **albedo**) than dark surfaces like grass or forest; and thick cloud like cumulonimbus has a higher albedo than thin cloud like cirrus.

Only about 24 per cent of the insolation arriving at the top of the atmosphere actually reaches the earth's surface directly; almost as much reaches the surface as diffuse radiation, which has been scattered on its passage through the atmosphere.

When the radiation reaches the ground most of it is absorbed, and converted into heat energy. Some is also used by plants for photosynthesis. Much of the heat energy is **reradiated** into the atmosphere, this time as **long wave radiation**. These rays are not able to pass through the atmosphere as easily as the incoming short wave rays did; 96 per cent of the rays are trapped, mainly by water vapour and carbon dioxide. This is the natural **greenhouse effect**, which maintains the earth's temperature as we know it.

There is a greater receipt of insolation in the equatorial regions than there is in polar regions. This leads to a **heat transfer** which moderates the temperature difference. Most of the heat (about 80 per cent) is moved polewards by winds, either at the surface or higher in the atmosphere. Some (20 per cent) is transferred in ocean currents. A second flow of heat transfers energy from the surface to the atmosphere. Conduction, convection and radiation are important means of allowing this flow, but so is the transfer of **latent heat** due to the evaporation and condensation of water.

Factors affecting temperature

The key factors affecting temperature are:

1 **Latitude** Places near the equator receive more concentrated insolation than places near the poles, because of the angle at which the sun's rays strike the surface. In addition, rays arriving here have had to pass through less atmosphere. Figure 4.1 shows why the equator is hotter than the pole.
2 **Altitude** Air at height is less dense than air close to sea level. Therefore it is able to hold less heat. This means that, on average, temperature falls by about 6.5 °C per 1 000 m.
3 **Distance from the sea** The sea has a greater specific heat than the land. This means that it takes more energy to raise its temperature than it does to raise the temperature of the land. As a result the sea heats up more slowly in summer, but it also retains its heat

Figure 4.1

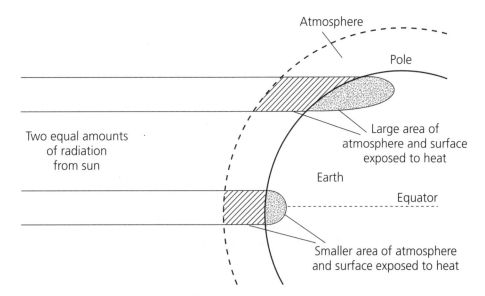

longer in winter. This effect is increased because the sun's rays can penetrate deeper into the sea than they can into land, so the sea is warmed to a greater depth.

4 **Prevailing winds** Winds take on the temperature and humidity characteristics of their areas of origin. They transfer those features to the area into which they blow. In general:
  – winds from the low latitudes bring warmer temperatures
  – winds from the sea are warm in winter and cool in summer

5 **Ocean currents** Also transfer large amounts of heat. As a rule ocean circulations are clockwise in the northern hemisphere and anticlockwise in the southern hemisphere. Currents to the east of land masses generally flow from the equator and bring higher temperatures; those on the west flow from the poles and bring cool conditions (although the North Atlantic Drift is an exception to this rule).

6 **Seasonal changes** These are due to the movement of the overhead sun.

7 **Length of day and night** This varies seasonally. It is a minor factor at the equator, where there is little variation throughout the year. Towards the poles it becomes far more significant, because of variations caused by the tilt of the earth's axis.

8 **Aspect** Locally the orientation of slopes can have an important effect on temperature. Insolation is more concentrated on slopes facing south in the northern hemisphere.

The humidity of the atmosphere

Air is a mixture of several gases. From the point of view of the meteorology student, water vapour is probably the most interesting. Humidity is a measure of the water vapour in the air. The amount of water vapour that can be held depends on the temperature of the air. As air warms up it is able to hold more water vapour.

Absolute humidity describes the amount of water vapour in a given volume of air, usually in grams/cubic metre (g/cu m).

Relative humidity (RH) is a more important measure.

$$RH = \frac{\text{the amount of water vapour held in a certain volume of air}}{\text{the amount that } \textit{could} \text{ be held in that air, at that temperature}} \times \frac{100}{1}$$

If air at, say, 90 per cent RH is cooled it will approach 100 per cent RH. When it reaches that point it is said to be **saturated**. The temperature at which it becomes saturated is called **dew point**. If it is cooled further it will be unable to hold all its water vapour, so some vapour will condense and turn back into liquid water.

The opposite happens when air at less than 100 per cent RH passes over a water surface. Then some of the water evaporates and turns into water vapour. Evaporation uses energy, and usually takes heat from the environment, which means that any surface from which water is evaporated is cooled. The energy needed is called latent heat. When condensation takes place the latent heat is released again, warming the air around it. This principle is important for an understanding of **adiabatic lapse rates**.

Causes of condensation

Water vapour condenses to form precipitation when the air is cooled below dew point. Cooling takes place in the following ways:

1 **Radiation** Occurs when heat is radiated from the ground, and air in contact with the cooled ground is also cooled. This is most likely to happen on calm, clear nights with anticyclone conditions. This often produces fog. If water is condensed onto the ground or leaves etc. it produces dew or frost if the temperature is below freezing.
2 **Advection** Happens when moist air moves over a cooler surface. It is common when air from a warm sea area drifts over a cold current. It also causes fog.
3 **Forced ascent** Occurs when air is forced to rise over high land (orographic or relief ascent) or over a colder, denser air mass (frontal ascent). As the air rises it expands, and energy is used up in the process of expansion. This causes a drop in temperature, and can lead to cloud formation, followed by precipitation in the form of rain, snow or sleet.
4 **Convective** or **adiabatic cooling** Is when pockets of air on the ground are warmed, expand, become less dense than the surrounding air, and start to rise. The air expands as it rises using energy causing heat loss. The air goes on rising as long as it is less dense than air. The height to which it rises depends on the nature of the surrounding air (the **environmental lapse rate** or ELR), and on whether the air reaches dew point. Dry air rises at the **dry adiabatic lapse rate** or DALR (1°C/100 m); but once it becomes saturated it rises at the **saturated adiabatic lapse rate** or SALR. This is associated with a slower rate of cooling than the DALR, and so the air rises faster, and goes on rising further once the SALR is reached. This sort of cooling also produces rain, snow and sleet, but often also produces hail. Thunder and lightning are often associated with this type of cooling.

Stable and unstable air

A precise knowledge of lapse rates is necessary for a full understanding of stable and unstable air. However, a basic answer on this topic should include the following points:

1 **Stability** Occurs when heated air rises, cools quickly and soon reaches the same temperature as the air around it. At this point it stops rising.
2 **Instability** Is when rising air cools, but stays warmer than the air in the environment through which it is rising. Consequently it remains less dense than its environment, and goes on rising. If such rising air reaches dew point, and starts to cool at the SALR, its rate of cooling will become slower. Therefore the difference between its temperature and that of its surroundings increases. The air rises still faster, producing high, thick, cumulus clouds often leading to heavy rain.
3 **Conditional instability** Is an intermediate stage. The air is stable, as long as it cools at the faster DALR. If the air is forced to rise so far that it reaches dew point then it becomes unstable.

Air masses and air streams

An **air mass** is a body of air which forms over a region which has uniform conditions of temperature and humidity. These conditions are passed upwards to the air mass. Temperature and humidity become the same throughout each layer of the air. When an air mass moves from its source region it carries certain characteristics with it, affecting the weather of places it moves to; but at the same time the lower layers of the air mass are modified by contact with a surface with different conditions. The table on p. 33 describes the main types of air mass which affect Britain and Western Europe.

When air moves from its source region it forms an **air stream**. This is a flow of air from high pressure to low. The strength of an air stream depends on the **pressure gradient**; when there is a big difference between the high and the low the wind will be strong, and a weak pressure difference will produce a gentle wind. However, winds do not blow straight from high to low, as they are deflected by the effects of the rotation of the earth. In the northern hemisphere winds blow out from high pressure regions in a clockwise direction, and into low pressure regions in an anticlockwise direction. In the southern hemisphere this is reversed.

Air mass	Source region	Temperature	Humidity	Effect of movement	Weather conditions
Arctic (A)	Arctic Ocean in winter.	Very cold.	Moderate absolute humidity, because cold air cannot hold a lot of vapour.	Moves south over warm sea. Lower layers warmed. Picks up some moisture. Becomes less stable.	Rare. Bitterly cold. Some snow and hail in heavy showers.
Polar maritime (Pm)	N Canada. N Atlantic. All year.	Cool – summer. Cold – winter.	Moderate.	Warms slightly as it passes south. Becomes unstable in lower layers.	Very common. Cold. Showers of rain or snow, separated by bright periods. Forms cold sector of depressions.
Polar continental (Pc)	Siberian high pressure in winter.	Cold.	Dry.	Stable, until it reaches the North Sea. Then warmed and picks up moisture. Becomes less stable.	Forms 'blocking anticyclone' and can last several days. Cold, crisp weather, but can bring snow especially to East England. High wind chill possible.
Tropical maritime (Tm)	Azores high pressure.	Warm – summer. Mild – winter.	Wet.	Lower layers cool as it moves north. Becomes more stable.	Mild in winter; warm in summer. Cloudy. Rainfall, if triggered by relief or front. Forms warm sector of depressions.
Tropical continental (Tc)	North Africa, SE Europe, in summer.	Hot.	Dry.	Lower layers cool as it moves north. Becomes very stable.	Not common, but brings heat wave conditions. Hot, dry, hazy. Thunderstorms caused by convective uplift.

Fronts and depressions

Despite the important influence of air masses and air streams, depressions are probably the single most important feature of the weather in Western Europe. They form along the **polar front** which is where the tropical air, moving from the southwest, meets polar air moving from the north. Waves form along this front, probably linked to waves in the upper atmosphere air flow or **jet stream.** These waves form over eastern N America or the north Atlantic when tropical maritime air pushes northwards into polar maritime air. This process forms an **embryo depression**, Figure 4.2, which starts to move from west to east along the polar front.

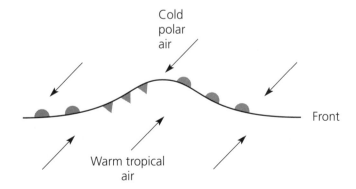

Figure 4.2

Cold polar air

Front

Warm tropical air

As the depression moves eastwards two separate fronts are formed. The **warm front** marks the arrival of the **warm sector**, and the **cold front** marks the arrival of the **cold sector** as shown in Figure 4.3.

Figure 4.3

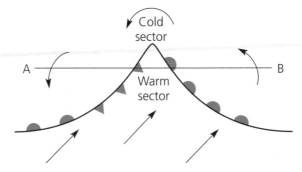

As the depression develops the air in the warm sector, which is less dense than the cold air, is gradually lifted off the ground by the denser cold air. The effect of this can be seen in cross-section (AB) on the diagram below. Cloud forms at both fronts, but it is usually more extensive at the warm front, which has a gentler gradient. Higher, thicker clouds, bringing short periods of heavy rain, form at the cold front. As depressions move from west to east, the sequence of weather during a depression's passage can be understood best if the following diagram is read from right to left. Over a period usually lasting several days the cold front catches up with the warm front, at ground level. This forms an **occluded front**. The air in the warm sector is lifted off the ground, as shown in Figure 4.5, and the depression decays away or **fills**.

Figure 4.4

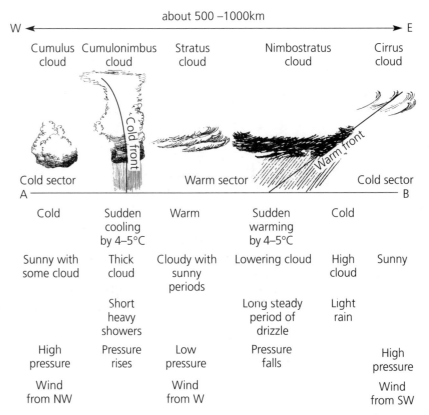

Figure 4.5

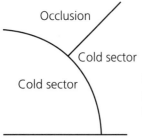

Tropical cyclones

In many ways cyclones are the tropical equivalent of the mid-latitude depressions. They form:

► between 5° and 20° north or south of the equator
► over seas with a temperature of at least 26 °C, heated to some depth
► in late summer or autumn, when the sea is at its hottest

The hot sea provides a source of energy and water vapour. Heated air rises, drawing in more air to be heated and raised. The rising air cools, causing condensation. This forms high cumulus clouds, and releases large amounts of latent heat, providing further energy for the system. Wind circulates into the low pressure thus formed, and the earth's rotation at these latitudes makes the whole system circulate at increasing speed. A cyclone of low pressure, strong winds, towering clouds, heavy rain and surging waves can reach 750 km in diameter. Cyclones move westwards, round the edge of the sub-tropical high, until they reach land. Here they cause devastation, but they rapidly decay over land as they are cut off from the sea – their source of energy and water vapour.

The three cell model

This is probably the aspect of meteorology which is the most important to help in understanding the pattern of world climate regions needed in climatology. It also helps with an understanding of the world's circulation system and wind belts. Figure 4.6 shows the tricellular model of global circulation.

1 The **Hadley cells** lie to the north and south of the equator. The area of maximum insolation is always close to the equator although it moves north and south with the seasons. Solar heating leads to convection currents of unstable warm air, which rise, and cool at the SALR. This produces towering clouds and the frequent torrential rainstorms which are characteristic of the equatorial regions. The rising air also leads to low pressure conditions forming at the surface. When this air reaches the **tropopause** it stops rising, and flows polewards.

   Much of the air sinks back towards the surface at about 30° N or S. As it sinks it is warmed, but this time at the DALR, because there is no moisture in the air to be evaporated. This produces regions of high pressure, where the air is hot and dry with cloudless skies. Winds blow out from the **sub-tropical high pressure** zones. Those flowing towards the equator are called the **trade winds**. The northeast trades and the southeast trades converge at the **Inter Tropical Convergence Zone** (ITCZ), where the air rises again, completing the cell.

2 The **Ferrel cell** lies to the north of the sub-tropical high. It receives its energy from the warm, westerly winds flowing polewards from the high pressure. They meet cold polar air along the north and south polar fronts. This forms the depressions (described

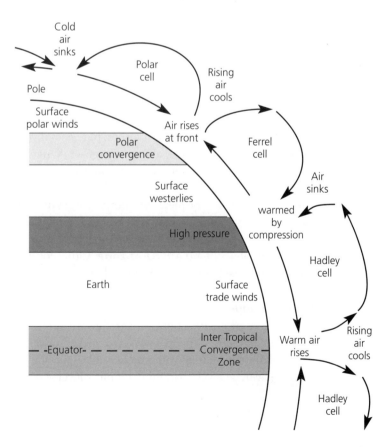

Figure 4.6
The tricellular model of global circulation

above) and the warm air is forced to rise and cool. Air then flows back towards the tropics, and converges with the high level air in the Hadley cells. This air also warms as it sinks.

3   The **polar cells** also receive some of the air which rises at the polar fronts. This air flows towards the poles, and finally sinks, producing the polar high pressure areas. Air flows out from these to the polar fronts. The polar cells are the weakest and least developed of the three cells.

Movement of the cells and the wind belts

So, in each hemisphere there are three distinct bands of surface winds, with their associated areas of high and low pressure. However, they are not still. It is the sun which provides the energy that makes the system work, and the overhead sun moves with the seasons. As a result the pressure and wind belts move too. Some areas lie in the same wind belt throughout the year; other places experience changes as the belts migrate, and these changes of wind and/or pressure can cause marked seasonal variations. Figure 4.7 shows the generalised movement of the wind belts.

Figure 4.7
Generalised movement
of the wind belts

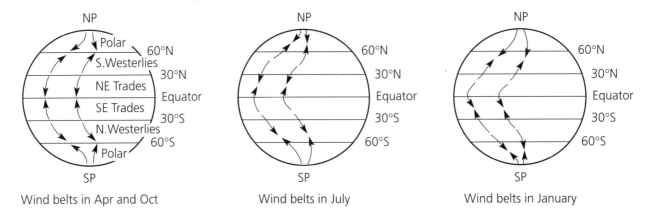

Wind belts in Apr and Oct   Wind belts in July   Wind belts in January

Climatology

In the space available here it is not possible to provide more than the briefest of outlines of what must be revised for an examination on climate. Different syllabuses emphasise different regions, and so the detailed content of your revision must be left as an individual decision. However, for any climate region studied you should, ideally, know the main defining statistics; but, even if you do not learn these statistics exactly, you should at least be able to recognise them when presented with them. For each region that you have to study you could complete a summary chart like the one on p. 37.

Of course there will be local variations within any region, and so you should add details of these variations where necessary. Often it is useful to distinguish between the core areas and the margins of your region, where your climate type starts to merge with its neighbours in a **transition zone**. Other local variations are due to differences between maritime and continental influences, altitude, etc.

Climatic change

This topic is topical, and controversial. Beware of topical topics! Many bad answers are written on them, by people who have picked up a little knowledge from the press, but who cannot back it up with precise understanding and factual detail. A little knowledge can be a very bad thing for an A-level geographer.

Climate type	
Location	
Highest temperature (°C)	
Season of highest temperature	
Lowest temperature (°C)	
Season of lowest temperature	
Total precipitation (mm)	
Seasonal distribution of precipitation	
Seasonal pattern of wind and pressure systems	

Climatic change has been happening since as long ago as evidence is available. It is not a new phenomenon. Evidence comes from:

▶ the geological record (ice ages are a particularly good source of evidence)
▶ changes of sea level
▶ pollen analysis
▶ tree dating
▶ the historical record

Suggested causes include:

▶ Changes in the output of energy from the sun.
▶ Cyclical changes in the earth's position in space:
  – the orbit round the sun may change shape
  – the tilt of the earth's axis may change angle
  – the earth may 'wobble' on its axis
▶ The ocean circulation may change, due to complex causes which are not fully understood. This can result in the 'El Nino' current in the Pacific.
▶ Movement of the plates may affect the relationship between land and sea areas.
▶ Meteorites and volcanic eruptions may throw dust into the atmosphere, blocking out the sun; and so on.

Human activity does almost certainly seem to be affecting climate, but the evidence is still not conclusive, and all answers should be expressed cautiously. The two most likely effects are:

▶ **The greenhouse effect** Where the release of gases into the atmosphere, especially carbon dioxide from the burning of fossil fuels, but also methane from cattle, CFCs from aerosols, etc., is trapping long wave radiation from the earth. This is probably causing heating, and may lead to melting of the ice caps, followed by a rise in global sea level.
▶ **Desertification** Caused by over farming in dry marginal regions. This may have led to reduction of vegetation cover, increased run-off and reduced evapo-transpiration, subsequently reducing precipitation. The whole process becomes a downward spiral, leading to disastrous results.

Urban micro-climate

This is an important topic on some syllabuses. Lack of space does not allow it to be considered fully here, but areas of study should include:

▶ differences in albedo or reflectiveness of heat
▶ urban heat islands
▶ the effect of cloud and dust in the atmosphere, reducing insolation, or trapping heat in
▶ the formation of photo-chemical smog
▶ wind and turbulence, affected by buildings
▶ variations in climate due to aspect of buildings
▶ changes in precipitation, caused by buildings' increased output of water vapour, or by clouds formed in urban thermals

★ REVISION ACTIVITIES

Complete the following sentences; then learn them precisely.

1 The **Environmental Lapse Rate** (ELR) is the fall of temperature experienced with_____.
It is usually about _____°C per 1 000 m, but can vary from place to place, and time to time.

2 **Adiabatic lapse rates** describe what happens when 'bits' of air r_____, e_____, and c_____; or when they s_____, become c_____, and w_____.

3 The **Dry Adiabatic Lapse Rate** (DALR) is the rate at which _____ _____. It is constant at _____°C per 1 000 m.

4 Rising air cools (and sinking air warms) at the **Saturated Adiabatic Lapse Rate** (SALR), when it cools to dew point. _____ is released as _____ _____. This means that the air cools _____ quickly than at the DALR. If air starts to cool at the SALR it will rise farther and faster. This will produce _____clouds, and _____.

5 Air is said to be **stable** if rising, unsaturated air cools _____ than the air in the environment around it. This means that the air usually does not reach d_____ p_____, and so cloud and rain _____, unless the air is forced to rise by _____ _____.

6 Air is **unstable** when rising air cools _____ than the air in the environment around it. When it reaches _____ it starts to cool even _____, at the __ ALR, because of the release of _____.

7 _____ occurs when the ELR is lower than the DALR but higher than the SALR. This means that the air will be stable, as long as it is cooling at the _____. If it becomes s_____ and starts to cool at the _____ it will rise freely, causing rain showers.

? PRACTICE QUESTIONS

Question 1

Table 1 overleaf shows monthly temperature and rainfall statistics for Plymouth, a coastal town in southwest England, and Oxford, a town situated in central England.

(a) Compare the temperature and rainfall statistics of the two towns and explain any differences you identify. (10)

(b) Discuss other climatic characteristics that you might expect to differ between the two towns. (8)

Table 1

Climate statistics for Plymouth

Month	J	F	M	A	M	J	J	A	S	O	N	D
Temperature (°C)	7	8	9	10	13	15	17	15	14	12	10	8
Rainfall (mm)	110	75	70	53	60	55	70	72	75	96	105	110

Total rainfall (mm) = 951

Climate statistics for Oxford

Month	J	F	M	A	M	J	J	A	S	O	N	D
Temperature (°C)	4	5	8	10	13	15	18	17	14	12	8	5
Rainfall (mm)	65	48	45	50	52	45	60	55	55	60	65	60

Total rainfall (mm) = 660

(c) With reference to specific examples, describe and explain how some plants are specially adapted to cope with the particular climatic conditions that prevail in coastal environments. (7)

(Oxford)

Question 2
Describe and explain the differences between temperate depressions and tropical cyclones (hurricanes). (20)

(London)

Question 3
Figure 4.8 shows unstable atmospheric conditions.

Figure 4.8

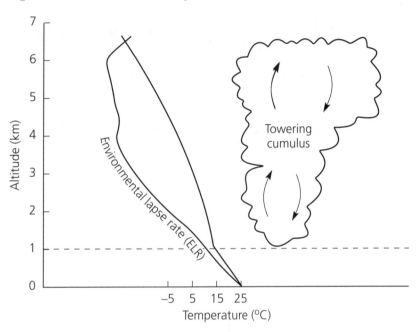

(a) Complete the diagram by labelling the following features:
  (i)   dry adiabatic lapse rate   DALR
  (ii) saturated adiabatic lapse rate   SALR
  (iii) condensation level   CL   (3)

(b) What is meant by the environmental lapse rate (ELR) shown on the diagram?
[2 lines] (2)

(c) Explain why the saturated adiabatic lapse rate (SALR) is at a different rate from the dry adiabatic lapse rate (DALR). [2 lines] (2)

(d) (i) Describe the weather conditions normally associated with unstable air conditions.
[4 lines] (4)

  (ii) Areas reliant upon tourism benefit from stable air conditions. Suggest *two* reasons why this is so. [7 lines] (4)

(AEB)

Question 4
(a) (i) Explain what is meant by the term *energy budget* in relation to the atmospheric system.
  (ii) Identify the roles of the different contributors to that budget. (8)

(b) Study Figure 4.9 and Figure 4.10 on p. 40 which show the location and mean monthly precipitation figures for a number of places in Africa.
  (i) Group these places according to different rainfall patterns.
  (ii) Justify your decisions. (7)

(c) To what extent is it justifiable to regard the monsoon climates of India and SE Asia as being distinctly different in nature from the climates found in other tropical regions of the world? (10)

(UCLES)

Figure 4.9

Figure 4.10

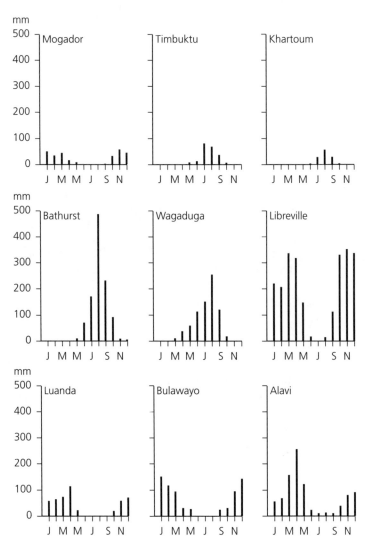

© I. J. J. Jackson, Climate, Water & Agriculture in the Tropics, adapted by permission of Longman Group Ltd., 1986.

Question 5

Consider the possible causes of global warming and suggest the likely climatic consequences.   (20)

(*WJEC*)

5 **Population and resources**

Population

Population numbers and density

The population of a country is measured by a census. In most countries censuses are carried out every ten years, often in years ending in 1 (e.g. 1991). In the economically more developed countries (EMDCs) they are usually very accurate. They are less reliable in economically less developed countries (ELDCs), because of problems of cost, organisation, illiteracy, accessibility, suspicion of the government in some places, etc.

Total population of a country or region is usually expressed as a plain figure. **Population density** is a very useful statistic, and is usually expressed in terms of people per square kilometre. Some small, highly urbanised states, such as Singapore, have a very dense population with average densities of over 5 000/sq km. Other countries, such as Saudi Arabia, have a very sparse population with fewer than 10/sq km.

Dense population should not be confused with **overpopulation**. This is an expression of the balance between people and resources. It is difficult to define overpopulation, but it is when the resources are inadequate to provide a reasonable standard of living for the population of the country (or region), given the present level of technology. Mention of level of technology is important because, for instance, an area might not be able to support all the people in a primitive farming community; but if farming technology improved people might raise their yields and be supported very adequately.

Underpopulation means that there are not enough people in a country to develop the resources to their full potential. In such a case an increase in population should allow greater and more efficient division of labour, leading to increased output per person, and raised living standards.

If overpopulation and underpopulation can exist, there must be a state of **optimum population** density, where the population is exactly right to develop the resources and achieve the best possible standard of living. Unfortunately this is even more difficult to recognise. At best the optimum is dynamic, varying as technology alters people's ability to use the resources.

Population change

The population of any country or region changes constantly. This can be as a result of natural change or migration. Natural change is due to births and deaths. The **birth rate** (BR) is the number of live births per thousand (‰) of the population, in any year. Britain has a low birth rate of about 10‰. Some ELDCs have high birth rates of over 40‰. **Death rate** (DR) is the number of deaths ‰ population, per year. Most EMDCs have low death rates, below 12 ‰. Most ELDCs have falling death rates. Natural population increase (or decrease) is the difference between BR and DR. It is usually expressed as a percentage, and care must be taken when transferring between BR and DR ‰ and Change %.

It should be noted that population increases are cumulative. For instance if the rate is 2 per cent in a place with 100 units at the start of year 1, it will have 102 units at the start of year 2, so its population will grow by 102 × 2 per cent in that second year. A 2 per cent growth rate will lead to a doubling of the population in about 35 years.

Two other important measures of population change are **infant mortality**, which is the percentage of children who die before they reach their fifth birthday, and **life expectancy**, which is the average age that the people live to.

Migration also affects the total population. **Immigration** is people coming into a country, and **emigration** is people leaving. Each is expressed as a percentage, which compares the number of arrivals or departures during a year, with the actual population at the start of the year. The population equation could be shown as:

Population at = Population at + (Births − Deaths) + (Immigrants − Emigrants)
end of year   start of year

In the past it seems that population growth was limited by the availability of resources within any region. If overpopulation occurred then the death rate increased, or emigration occurred, and the population fell back to reach equilibrium with resources. High population density could only occur in resource-rich areas. However, in the twentieth century rapid communications allowed medical advances to spread quickly. Along with transfer of food supplies, etc., this helped to keep people alive, even when the region's natural resources were inadequate, or underdeveloped.

Malthus gave his name to a school of thought which said that population was likely to increase faster than food supply, unless strenuous efforts were made to control the birth rate. He made his predictions in the late eighteenth century, but they have not come true yet, because technological development has allowed agriculture to become ever more efficient. However, in the late twentieth century **neo-Malthusians** have suggested that his predictions are about to be fulfilled now, especially in ELDCs.

The demographic transition

In the 1960s it was realised that many ELDCs were undergoing a massive population increase, which was threatening to bring famine and disaster on an enormous scale. Overpopulation was clearly occurring. Study of countries in Western Europe showed that many of them have gone through four or five stages of population change, as they developed. This was the basis of the **demographic transition model**. It seemed that many ELDCs had gone through the first two or three stages, and the disaster could best be averted by ensuring that they reached the fourth stage as soon as possible. The stages of this model are shown in the diagram below.

Figure 5.1
The demographic transition model

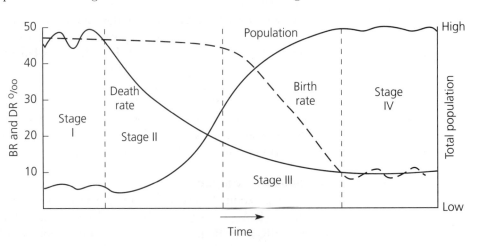

In stage I technology and communications are poor, the death rate is high, and the birth rate has to be correspondingly high to ensure maintenance of population. Though population fluctuates it is in a state of equilibrium with resources.

In stage II death rate falls with the introduction of technological advances in medicine, hygiene, water supply, transport, etc. Birth rates stay high, usually because children are seen as being necessary for economic development and security, especially in rural areas, and the costs of rearing children are low.

In stage III death rate continues to fall, and now birth rate starts to fall. This may be partly due to improved methods of, and access to, contraception. However, this alone will not cause birth rate to fall, unless individuals want to limit their family size. It seems that this is most likely to happen when people are secure enough to know that their children will all probably survive, and that they will be able to support themselves without needing large families. It seems that economic development for the majority of the population has to precede the fall in the birth rate.

In stage IV BR and DR are approximately equal. Population and resources are usually in equilibrium again, but at a high stage of technological development, so that the resources can support a dense population. If the BR falls below the DR then problems of an **ageing population** might arise.

Some demographers have identified a stage V, in which population starts to decline. In fact this might be inevitable, because ageing populations will inevitably lead to a rise in DR as a large proportion of people reach the natural limits of their life span.

Population structure

As the old joke goes, population structure shows the whole population broken down by age and sex. The most common way that it is shown is in **population pyramids**. Five 'sketch pyramids' have been given in the following diagram, to show the four stages of the demographic transition (and the possible fifth stage). You must know how to construct an accurate pyramid, but you will never be asked to do this in an examination. Instead you may find it very useful to be able to draw such sketches quickly in an examination, so these should be learnt.

Figure 5.2
Sketch population pyramids to match the demographic transition model

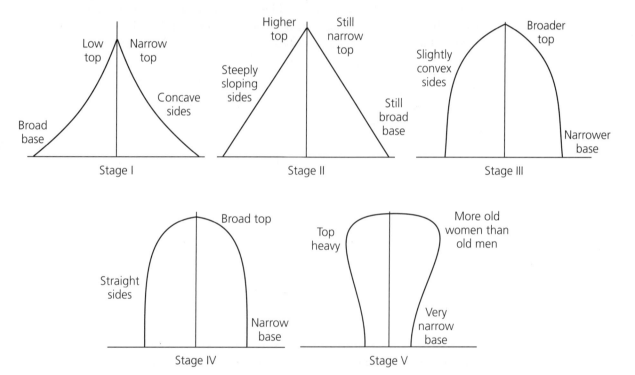

The **dependency ratio** shows the proportion of the population which is in non-productive age groups (below 16 and over 60/65) compared with the number of people in the potential workforce.

Migration

Demographers see migration as a way of equalling out imbalances between population and resources. If free movement of population was allowed people would move from areas of overpopulation to underpopulated places. However, governments usually put up barriers to discourage free migration and to preserve resources for the people who are already there. Exceptions are made:

► when people with skills are needed to help the country develop
► when shortage of population means that an increased labour supply is needed

It is possible to identify major movements of groups of migrants. Some of these are a result of **forced migration** but in most cases migration is **voluntary**. Such migrants move as a result of individual decision making. The **push/pull** model explains how these decisions are made. Pushes are negative features making people want to leave, and pulls are the positive attractions of the destination. Some of the pulls may be illusory pulls but they still have a big influence on decision making. The pushes and pulls have to be strong enough to overcome **anchors** or ties binding a person to the original home, and **barriers** which are things which make moving difficult.

Migration can be classified as permanent or temporary (which can be sub-divided into semi-permanent, seasonal or daily); forced or voluntary; internal (from one region of a country to another) or international; and based on mainly economic or social factors. In ELDCs the main flow of migration in recent years has been rural–urban or **urbanisation**; whilst in EMDCs urban–rural migration, or **counter-urbanisation** has been dominant. Preparation for the examination should include learning examples of each type, and then this classification, along with the push/pull model, can provide a basic essay structure.

What are resources?

It should be obvious from what has been written above that the study of population cannot be separated from the study of resources. Resources are those things that are needed by people. Something only becomes a resource when technology develops so that the thing actually becomes usable. Resources can be divided into human and physical resources.

Human resources
People provide power through their work, but also provide skills. The skill level of a population can be developed through education (for general development) and training (to do specific jobs). Through producing more than they consume, workers can produce **capital**. Capital represents stored production, or wealth. This can be concentrated in the hands of a few individuals, or can be shared widely, or it can be owned by the state on behalf of all the people.

Capital is often represented by money, and is the most mobile element of production. As communications improve, international movement of capital is having profound effects on the location of all types of economic activity. Industry is becoming far more mobile, as the owners of capital transfer production to the region or country where other resources, especially labour, are most easily (and cheaply) available.

Physical resources
These come from the atmosphere, plants and animals, sea, soil and rocks. They can be sub-divided into **renewable** and **non-renewable** resources.

Some renewable resources come in a constant flow (e.g. solar energy, tidal power, river flow), but others are more precisely renewable (such as plant crops which are renewed each year, and soil which is renewed as rock weathers). Such renewable resources have to be very carefully managed.

Non-renewable resources are those that cannot be replaced in the foreseeable future, because they take so long to form. They are mainly mineral resources. Some of these are recyclable (e.g. metal ores, glass from sand); but others, especially fossil fuels, are non-recyclable. Use of the non-renewable resources often releases matter into the environment in concentrations which cause pollution.

The changing location of resource exploitation
When resources are in plentiful supply, compared to the demand for them, they are cheap. Only the most easily accessible resources are used. As the accessible resources are used up the price rises, and the supply falls. Rising demand can also cause price increases.

As the supply is reduced, and price rises, this will reduce the rate of consumption. It also means that resources that were previously considered to be uneconomic, or too difficult to exploit, will now become worth exploiting. Rising prices also lead to the search for alternative resources becoming worthwhile.

Such **supply and demand mechanisms** have very important consequences for the pattern of exploitation of minerals and other resources. The changing location of the sources of minerals such as oil can be explained by reference to the mechanisms described above.

When people talk about resources 'running out' this is rarely strictly true. Rather, as a resource becomes rarer its price rises, through the action of supply and demand. Either fewer people can afford to use the resource; or the higher price stimulates people to search for new sources; or they are stimulated to find new ways of producing the resource. If none of those things happens, and the resource becomes too expensive to exploit, then what remains is abandoned.

REVISION ACTIVITIES

1 Write a definition of each of the following words and phrases – none should be more than one or two sentences long – then learn each definition.

census	population density
optimum population	natural population change
birth rate	death rate
infant mortality	life expectancy
immigration	emigration
push and pull	barrier and anchor
urbanisation	counter–urbanisation
population structure	dependency ratio
human resources	physical resources
renewable resource	constant flow resource

2 Complete the table below describing the demographic transition and then answer the following questions. Use single words or short phrases to fill in each section, thereby making it easy to learn and reproduce for the examination.

	Stage I	Stage II	Stage III	Stage IV
Birth rate				
Death rate				
Population change				
Total population				

(a) What defines the change from:
  (i)   Stage I to Stage II
  (ii) Stage II to Stage III
  (iii) Stage III to Stage IV

(b) What causes the change from:
  (i)   Stage I to Stage II
  (ii) Stage II to Stage III
  (iii) Stage III to Stage IV

PRACTICE QUESTIONS

Question 1

(a) Outline the stages of the demographic transition model.   (8)

(b) With reference to your chosen case study country, examine the relationship between economic development and stages in the model.   (12)

(London)

Question 2
(a) Outline and critically discuss the relationship between population and food resources presented by Malthus. (12)

(b) Discuss the environmental consequences and conflicts arising from agricultural intensification in temperate countries such as Britain. (13)

(*Oxford*)

Question 3
(a) Explain why an optimum population is seldom attained. (10)

(b) Evaluate the policies used by a selection of countries to reach a balance between population and resources. (15)

(*AEB*)

Question 4
Study Figure 5.3 which shows population pyramids for two countries.

Figure 5.3

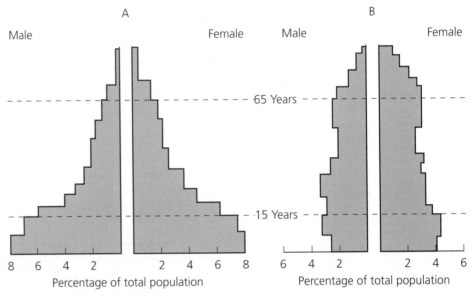

Percentage of total population		Percentage of total population	
Under 15 years	46	Under 15 years	28
Over 64 years	8	Over 64 years	19

(a) Calculate the dependency ratio for:
Country A [2 lines]
Country B [2 lines] (2)

(b) Identify and explain the differences between the two pyramids for the 15–39 age groups. [5 lines] (4)

(c) (i) With reference to a country or region you have studied, suggest *two* future problems which are likely to occur as a consequence of the population structure of the 0–15 year groups. [5 lines] (4)
(ii) How might each of these problems be overcome? [7 lines] (6)

(d) Comment on the social and economic consequences of an increasing number of people over 65 years of age. [5 lines] (4)

(*London*)

Question 5
With reference to specific examples, assess the extent to which internal migration has influenced the present distribution and density of population in Brazil. (20)

(*WJEC*)

6 Settlement

Site, situation, function and morphology

Studies of settlement come under four main headings:

- **Site** factors, which explain why a settlement was built in that particular place.
- **Situation** which examines how the settlement fits in to its region, and how it is linked and associated with other surrounding places.
- **Functions** which are the reasons for the settlement's existence, or the way in which people who live there make a living.
- **Morphology** which is a description and explanation of the shape and structure of the settlement.

All these factors must be viewed in the context of change over time. Settlements respond to changes in physical, social and economic circumstances, but each new development has to build on what has gone before, and almost all settlements show the influences of historical events in their present structures and functions.

Site factors

In order to analyse site you must ask yourself 'Why was this settlement built here, and not somewhere else a kilometre or so away?' The reason is usually linked to the physical nature of the land on which the earliest settlement was built. It may be to do with:

- flat land for ease of building, or for easy transport
- steep land for good defence
- near a river for water supply, or fishing, or transport
- at a route centre such as a bridge point, port, gap in a range of hills, or other nodal point
- safe from flooding, severe weather or other hazards
- at a resource location such as mine, power site, etc.

For any settlement studied you ought to be able to write, briefly at least, about its site. You should also have examples of settlements built at each of the above types of site. Note that the siting of a settlement is often influenced by more than one of the factors.

The situation

This involves a study of the relationship between a settlement and its region. Settlements provide services for the regions around them; and they also receive goods from the same regions. The development and prosperity of the settlement depends on the size of its area of influence, or its **hinterland** with which it trades. It also depends on the efficiency with which it can communicate with the surrounding region. The settlement may have a **core** sphere of influence, for which it is the major centre, and a **marginal region** beyond this, within which it shares its influence with other neighbouring or competing settlements.

Revision for the examination should involve learning how a settlement links with its surrounding area. It is often valuable to learn a sketch map showing the routes which converge on the settlement, the areas or settlements linked by these routes, the main flows of goods and services along these routes, and the negative factors which limit the influence of the settlement.

The relationships between settlements and their surroundings can be shown using a number of models. In the past the study of these models was a very important part of all settlement geography syllabuses. Now it is not always so important. All candidates are advised to be quite sure which are required for their syllabus, and then to understand and learn them as thoroughly as possible. Though they often appear complicated at first sight, they do provide clear structures on which to base examination answers.

The **nearest neighbour** model is a way of analysing the spatial patterns made by all the settlements in a region or country. It measures how clustered or dispersed they are.

The **rank–size rule** suggests that there is a numerical relationship between the settlements within an area or country. The rule suggests that if you know the size of the biggest settlement you can predict the size of all the other settlements. The rule often fails to work well, but it is useful because countries can be grouped depending on how they deviate from the rule. Primate distribution (with one city that has grown much larger than all the rest), and binary distribution (with two large cities of almost equal size) are the two most common deviations.

Central place theory says that settlements can be arranged into a hierarchy of size and functions. Larger settlements are fewer in number, have more functions with more specialised services, and larger hinterlands. **Christaller's central place model** was an attempt to predict how settlements would be related and spaced out in a country. It is a very useful model in some situations, but rather complex. You should take clear advice about how much knowledge of this model is needed for your syllabus. Understand and learn what is essential, but do not waste time learning what is unnecessary!

Gravity models show how places interact. Big settlements that are close together will have a lot of connections. As size decreases, or distance increases, nteractions will become fewer. **Reilly's law** of retail gravitation is a way of redicting where the break point between the spheres of influence of neighbouring towns will occur.

Functions of settlements

All settlements, except perhaps the very smallest, have a variety of functions. All, except perhaps the most recent, have undergone changes of function. However, it is useful to try to classify the main functions into groups. This certainly makes learning for the examination easier, and can provide a useful structure to help plan examination answers.

In rural areas the main functions of settlements have traditionally been to provide agricultural services, including a market for produce, and so they became transport centres. They also provided a place for defence in times of war. Provision of other services, including shops, banks, etc. gradually developed in many places. In recent years, in EMDCs, more and more rural settlements have become dormitory or overspill housing areas, for nearby cities. They have sometimes developed a leisure or tourist function, often becoming second or weekend home areas.

In urban areas many of the above functions still exist, although agricultural services are usually less important. Mining and manufacturing are often important. Provision of services, especially retail and wholesale, often dominates the Central Business Districts (CBDs) of towns and cities. From this trade develops a whole range of transport, banking, commerce, insurance and finance services. Then administration and government develop in some larger centres. Religious functions often come at quite an early stage in the settlement's development; other cultural functions, such as entertainment, sport, press and media then develop in some settlements. Resort functions are found in places with special attractions of scenery or history and culture. Of course the residential function is still important in all towns and cities.

Some writers try to make a distinction between the functions of cities in EMDCs and ELDCs, but this is a rather artificial division, which is becoming ever more blurred. Major cities across the world share all of these functions, although the proportions vary as cities and countries become more developed.

In a settlement where there is a variety of functions there is often competition between them for the sites with the best locations. Many functions wish to locate in the central parts of cities, where access is best. This competition has major implications for the structure of cities. So does the tendency for many providers of similar functions to need to locate near to each other, leading to clustering of functions.

Settlement morphology

Rural settlements range from individual houses, through hamlets to villages. Settlement patterns can be:

▶ **Isolated** Where individual buildings are separated by large areas of unsettled land.

▶ **Dispersed** Where individual houses or small hamlets are scattered across an area. This is typical of parts of Europe where the Celtic influence was strong.

▶ **Nucleated** Where buildings are grouped closely together. They may be centred around a route junction, a defensive point, a church, etc. Nucleated villages are usually surrounded by farmland. This form is often seen in English villages where Celtic influence was replaced by Saxon and then Norman influence.

▶ **Linear** Strung out, along a road, a river or valley, the edge of a flood plain, or a coastline. Such villages are often called ribbon villages.

▶ **Ring** or **green** Villages are grouped round an open central area. This may provide space for a market, for keeping cattle at night, or for social activities – including sports, e.g. the village cricket pitch.

Many villages actually have composite structures, with elements of more than one of the above patterns. A village may be nucleated, but have recent linear development along the road to the nearby town. In the EMDCs commuter villages have developed, where the old pattern has been almost completely obliterated by the growth of suburban estates around the edges.

Rural settlements only have a limited range and number of functions, and so they do not develop separate functional zones. However, the important functions tend to be concentrated at the centre. The village with the church, pub, and post office with general store all concentrated on the edge of the green is a common site in many parts of the UK.

Urban settlements have a far more developed structure, with clear differentiation of **functional zones**. These develop naturally, based on the idea of **bid-rent theory**. Planners also attempt to zone land use, and often try to devise urban structures which take things other than commercial factors into account. They try to make their cities better places to live, whereas commercial pressures are to make cities more profitable.

All cities have a **Central Business District (CBD)**. This is where the functions which can afford to pay the highest rents are concentrated. Most cities have **industrial zones**. These include manufacturing industry, and distributive industries. In the past these were often close to the centre, because they needed to be close to railways, which had few stopping-off points. Industry needed to be close to the terminals of these networks. In other cases the industry developed along axes such as provided by rivers or canals. In the late twentieth century, road transport has come to dominate. It has far more flexible networks, and so industry has moved away from the congested central locations, out towards the edges of cities.

Residential areas are also zoned. Older housing tends to be found towards the centre, although the oldest housing areas have often been cleared and redeveloped. The newest housing is usually built on, or beyond, the city limits. As well as being differentiated by age, housing is also found in distinct cost groupings. In the past the best housing was located in areas with good views, dry sites, open space nearby, clean air, or easy access to transport routes. The poorest housing tended to be built on land that was too damp, too polluted, too congested or otherwise unattractive to the rich. Once established, these cost and social divisions have tended to be perpetuated, with high cost housing being attracted to 'good areas', and low cost housing being built on cheaper land in 'less desirable' areas.

It is essential that students should know and understand the main models of urban development:

▶ **Burgess** Developed his model, based on the idea of residential zones, outward growth, and a succession of occupancy. This model has concentric rings of housing type as the major feature.

▶ **Hoyt** Built on the Burgess model, but added in distinctive functional zones. Industry developed in sectors along transport routes, and housing patterns were influenced by the location of industry.

▶ **Ullman and Harris** Developed the **multiple nuclei** model, which recognised that cities may grow from a number of different points. Each nucleus could become the focus of a quite distinct type of function.

▶ **Mann** Combined the ideas of the other models, especially Burgess and Hoyt, and applied them to British cities.

▶ **Third World Models** Grew from the Burgess tradition, but tended to reverse the pattern of housing zones. They placed the better class housing close to the centre, for easy access, with slum housing being pushed to the outskirts, on the cheapest land.

The details of all these models must be learnt thoroughly, but their practical applications should also be known. Candidates may be asked to comment on how well towns that they have studied fit with one or more of the models. You ought to know where named locations within specific cities would fit on the models. If you are asked to criticise the models do not simply condemn them as unrealistic. The models were intended as generalisations, not as perfect fits with any one city. Answers must acknowledge that the models all show some useful truths, and provide a check against which real examples can be compared.

The problems of cities

A large and growing proportion of the world's population lives in cities. They present enormous attractions to people of all ages, classes, levels of education and wealth. They also cause certain problems. In fact some cities may be said to be victims of their own success.

It is unfortunate, but geography courses often seem to emphasise the problems of cities, at the expense of the opportunities provided. You need to be aware of the problems, and should be able to give detailed examples for some of them. Then you ought to be able to describe how people and planners are attempting to tackle the problems, and again you should quote specific examples from one or more cities. Some problems and solutions are listed below for cities in ELDCs. Develop this list by adding your own examples, with references to case studies, and make a similar list for cities in EMDCs.

Problems of cities in ELDCs include:	Solutions to these problems include:
• **Housing** – there is not enough, so people are forced to live on the street, or in makeshift shelters. - much is badly constructed, from poor materials, on inadequate sites, on land which people do not own.	• Build high-rise flats (expensive). • Upgrade shanty houses. • Provide sites and/or materials for self-build schemes. • Negotiate with landlords for rights in return for rents.
• **Amenities** – many housing areas do not have running water, electricity, sewage, proper roads, etc.	• 'Site and service' schemes. • Provide septic tanks and regular collection of waste (cheaper than sewers).
• **Pollution** – drinking water is often contaminated, causing widespread disease. - industry is poorly regulated, so often pollutes with its waste products. - traffic causes air pollution, and high stress levels.	• Provide communal taps for communities. • Encourage recycling to increase profits. • Try to make deals to ensure that MNCs apply same standards in ELDCs as in EMDCs. • Invest in public transport. • Try to decentralise employment.
• **Employment** – new migrants are attracted by the hope of jobs, but unemployment, underemployment, and exploitation of the workforce are all common.	• Encourage MNCs to invest in labour intensive industries. • Support for the 'informal' sector. • Pressure from rich consumers to ensure products are fairly manufactured.
• **Transport** – public transport is often inadequate, poorly maintained and overused. – road networks are poorly developed outside the central areas. In the centres they are very crowded, with a mixture of cars, lorries and traditional transport, like rickshaws and bullock carts.	• Investment in public transport infrastructure, and subsidised fares, to encourage use, and reduce cars. • Investment needed. • Decentralisation – move jobs to the people.

Given this list of problems, it is not surprising that many candidates write answers which seem to suggest that cities in ELDCs, and especially their **spontaneous settlements**, are places of nightmare conditions. It must be appreciated, though, that people choose to come and live in these settlements. They come because the cities offer opportunities to make a living and to improve people's lifestyles. The priority of the residents is to make a living; it is not comfort; that will, it is hoped, come later. Some people have even gone so far as to describe some of the shanty towns as 'slums of hope'. Here, a majority of the inhabitants are benefiting from the opportunities of the city and improving their conditions. By contrast there are also 'slums of despair', where progress does not seem to be occurring

REVISION ACTIVITY

Consider a city with which you are very familiar. This will probably be your own nearest city, or it may be one about which you have done a very detailed case study.

(a) Name examples of as many of these functional zones as you can. Locate as many of them as possible on a sketch map.

▶ **Services CBD** – main shopping area; government administration area; large company administration area; ecclesiastical area (cathedral); university and education; hospital and health; main entertainment and leisure area; 'twilight' entertainment and leisure area; periphery of CBD (solicitors, dentists, small firms' offices, etc.). Peripheral Business Districts (office developments, etc. in the suburbs.)

▶ **Housing** – pre-industrial revolution; 19th century workers; 19th century middle class; inter-war workers; inter-war middle class - suburban/ribbon development; post-war council housing – inner city/outer city estates; post-war middle class; commuter villages; etc.
Areas of gentrification and inner urban redevelopment.

▶ **Industry** – surviving nineteenth century; nineteenth century derelict; twentieth century along roads to city centre; industrial estates; business parks; science parks; inner urban redevelopment; etc.

(b) Add some main transport features – railways, motorways, airport, port, etc.

(c) Add major areas of open space, parks, river flood plains, steep slopes, etc.

(d) On your finished map you should look for evidence showing where the Burgess, Hoyt, and Harris and Ullman models either match your town, or do not match it.

(e) Be prepared to reproduce the main features of your sketch map and analysis in the examination.

Note: answers to this revision activity have not been provided.

PRACTICE QUESTIONS

Question 1

(a) Identify *two* sources of evidence which might be used by the historical geographer to study the growth of urban settlements and describe how one of them gives an insight into this growth. (5)

(b) With reference to specific examples within England and Wales, discuss ways in which the political, social and economic processes operating in the pre-industrial town are reflected in the *present* townscape. (10)

(c) Discuss the possible issues which may arise within the modern-day urban environment as a result of the desire to preserve/conserve the pre-industrial elements of the townscape. (10)

(NEAB)

Question 2
Describe and explain the changing functions of rural settlements in Britain since 1960. Illustrate your answer by reference to specific examples.   (25)

(*Oxford*)

Question 3
With reference to *either* a large town *or* a city, identify the factors which cause its structure to differ from the models developed by Burgess, Hoyt, and Harris and Ullman.   (25)

(*AEB*)

Question 4
To what extent do you agree that both social and ethnic segregation within cities is due to economic factors?   (20)

(*WJEC*)

Question 5
Study the model of the structure of a hypothetical British city.

Figure 6.1

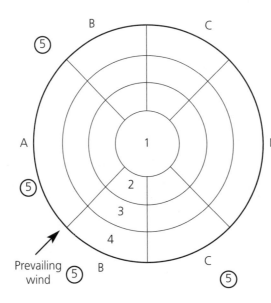

1 Central Business District
2 Transitional zone
3 Zone of smaller terrace houses
  in sectors C, D.
  Larger by-law housing in sector B.
  Large old houses in sector A.
4 Post-1918 residential areas with
  post-1945 development mainly
  on the periphery.
5 Commuting 'dormitory' towns.

A Middle class sector.
B Lower middle class sectors.
C Working class sectors (and
  council estates).
D Industry and lowest class
  housing sector.

(a) Give *two* reasons to account for the concentric arrangement of land-use zones 1–4 shown by the model.   [5 lines] (4)

(b) (i) Describe ways in which the type of housing in zone 4 might differ between sectors A and D.   [3 lines] (2)
    (ii) Suggest why these differences occur.   [5 lines] (4)

(c) For sector D, explain why industry nearer the inner areas might differ from that in the outer areas.   [6 lines] (5)

(d) Draw a simplified map of a *named* urban area you have studied, which identifies the different residential zones.   [Space for map. No lines] (5)

(*London*)

Question 6
(a) Study the table, overleaf, which shows the actual and projected incidence of poverty in four world regions.
    The Poverty Level is defined as: 'to be below the specific minimum level of income (assessed at $370(US) per head in 1985) needed to satisfy the basic physical needs of food, clothing and shelter in order to ensure continued survival.'

[*Source*: UNCTAD]

	Urban population below poverty level (%)		Urban population below poverty level (millions)	
	1985	2000 (est.)	1985	2000 (est.)
Asia (excluding China)	23	38	136.5	200
Africa (Sub-Sahara)	42	66	55.5	108
Europe/ Mid East/	34	28	59.5	68
Latin America	27	38	77.3	112

Suggest reasons for the differences shown in the table. (9)

(b) With reference to a *range* of urban areas you have studied:
   (i) Explain why growing numbers of the world's poor are found in urban areas.
   (ii) Outline some of the schemes developed to improve the quality of life of the
       urban poor. (16)

(*London*)

7 *Employment*

This section covers a very broad area of geography. The different syllabuses deal with the topics covered here in a variety of different ways, linking them together in different combinations. The structure used here tries to simplify the complexity, but take care that *your* revision fits *your* syllabus.

Employment can be divided into three (or possibly four) major categories:

► **Primary** Workers produce raw materials from the land. Primary employment includes farming, forestry, fishing, mining and quarrying. Economically less developed societies are characterised by high proportions of the population being dependent on primary production.
► **Secondary** Also known as manufacturing industry. Here the raw materials are processed to make finished products.
► **Tertiary** Provides services. These can be services to industry, or private services. They include transport, education, retail, leisure and entertainment, administration, military and emergency services, and many others.
► **Quaternary** These jobs have recently been defined as those to do with high technology and information. This is the fastest growing sector of many developed economies.

Primary employment

Some of the most important aspects of mining have been dealt with in Chapter 5. Here it is intended to deal with farming and agriculture.

All farm systems can be studied in terms of inputs, processes and outputs. The farm operates within a system of **physical controls** which limit what can be produced, but which present certain opportunities to the farmer. The farmer then has to exercise choice from among the opportunities.

Physical controls include:

► **Climate** Rainfall, sunlight and heat are essential inputs for plant growth, and they assist soil development. Wind, frost, etc. limit possibilities for plant growth.
► **Soil** Which is renewable, but which has to be carefully managed.
► **Slope, altitude** and **aspect** All influence climate and soil, but they also have a direct influence on what can be done by the farmer.

Economic influences are also important. Farmers must serve the market, unless they operate in a subsistence economy. However, in many areas, like the EU, government policy can alter market forces by providing:

► **subsidies, guaranteed prices**, and other incentives
► **tariffs** and other forms of protection from outside competition
► **quotas**, and other limits on production

The availability of capital is a factor of great importance. Where it is readily available, farms can be **capital intensive**. Where it is not, they are either **labour intensive** or **extensive**. Capital availability also influences access to technology and transport, which further influence the type of farming.

Human influences are closely linked with economic factors. They include the size of the labour force and its level of education and training. The farmer's aspirations should also be considered. Some farmers follow an optimiser strategy, aiming to produce as much as they possibly can under the conditions that are present. Others are satisficers, content with a certain standard of living who do not aim to develop further. Satisficers work in a more traditional way, whilst optimisers aim for modern development.

The systems that arise from these constraints and opportunities can be classified into the following categories:

1 **Arable** which grows mainly crops, especially cereals.
   Pastoral which mainly keeps animals.
   Mixed with a combination of crops and animals.
2 **Subsistence** where production is mainly aimed at feeding the farmer and her/his family; (note – 'her' before 'his' is not political correctness. In subsistence farming communities women usually do the majority of the farming work!).
   Commercial where production is for sale at the market. This includes plantation agriculture, where land – usually in ELDCs – is owned by large firms – usually from EMDCs – which farm the land on an industrial scale, with large inputs of labour and capital. They usually produce for export.
3 **Shifting** where the farmer moves from place to place, either with herds of animals, (usually seasonal movement – nomadism) or for new crop land (usually on a longer cycle).
   Sedentary which comprises the majority of farms which are permanently occupied.
4 **Extensive** where the farm covers a large area, and inputs of labour and capital per hectare are low.
   Intensive where the farm is usually smaller, and inputs per hectare are high. There are capital intensive and labour intensive farms, although some farms are both.

In preparing for the examination it may be useful develop a matrix of farm types, with learnt examples to fit the various combinations of factors.

It is essential that some of the problems associated with agriculture are clearly understood. These include:

▶ pollution caused by addition of chemicals as farming becomes more capital intensive
▶ loss of wildlife habitat, due to intensification, and the spread of agriculture into new areas
▶ soil erosion, which can be caused by removal of hedgerows, use of slopes that are too steep, over farming and destruction of soil structure, etc.
▶ desertification, also often resulting from inappropriate farming methods in marginal areas; it is also linked to climate change, whether as a cause or an effect is not always clear
▶ loss of species diversity, as hybridisation leads to domination of seed supply by agribusiness firms
▶ rural depopulation, caused by mechanisation reducing the workforce, and by amalgamation of holdings depriving many poor farmers of their land
▶ the takeover of subsistence land by commercial producers, which deprives many ELDCs of food crops, in order to produce export crops often aimed at the luxury market
▶ overproduction caused by subsidies and market intervention; etc.

Secondary employment

Before the industrial revolution all manufacturing was small scale, craft based, used simple sources of power, and few machines or tools. The industrial revolution harnessed coal power to drive machinery, which meant that industry could take place in factories, on a much larger scale. It became far more capital intensive. Technological development has meant that machinery has replaced much labour in industry.

The factors of industrial location include:

▶ **Power** Which was very immobile in the form of water power or coal, but is more mobile in the form of oil or gas, and very mobile as electricity. Therefore power is less dominant in location decisions than it once was.
▶ **Raw materials** Which are the products of primary industry, and components or parts, which have been produced by another factory. Their importance as a location factor depends on their bulk, fragility, etc. which affect raw material transport costs. When transport makes up a large proportion of the cost of the finished product the industry is drawn to raw material locations. In recent years transport has become increasingly efficient, and this has reduced the pull of raw materials.

▶ **Labour supply** Can influence location either when a large workforce, or a very specialised workforce is needed. In the former case industry is drawn towards areas of cheap labour, high unemployment, or low levels of economic development. In the latter case it is attracted to areas of high education and training. Some specialised industries locate in places with an environment which attracts the right kind of workforce.

▶ **Market** This can be the same as the labour force, especially in a large conurbation. Some consumer industries produce goods which everyone buys. Others produce specialised goods for a small part of the population. Some industries sell all their product to other industries. This might all go to a single firm, or to many. The type of outlet has a big influence on location. Perishable goods, including high fashion goods, traditionally located as close as possible to their market; although again, improved transport efficiency has reduced this link.

▶ **Space, environment, government policies** (which often means attracting industry to areas of high unemployment or dereliction) and **capital** are other factors. Capital in the form of money for investment is probably the most mobile of all the factors of location. The deregulation of the world financial system in recent years has made it even freer to move across national borders. Fixed capital, in the form of factories, machinery, etc. is fairly immobile; it often leads to **industrial inertia**, when a factory survives in an area long after its initial reason for locating there is no longer valid.

Learn these factors with reference to specific examples to illustrate each one.

The location quotient

The Location Quotient (LQ) is a useful statistic for the geographer. It shows how much of an industry is concentrated in an area. The national average LQ for any industry is always 1.0. If the LQ for the industry in an area is more than 1 it means that the industry is more concentrated in that area than it is in the whole country; if the LQ is less than 1 the industry is less concentrated there than in the whole country.

The figure is so useful because it shows spatial distributions very clearly. The formula is quite complicated, but, at least, you should understand how the final LQ results can be interpreted.

Changes in industrial location

In 1909 **Weber** produced his model of industrial location. This was based on the least cost of transport. It was ideally suited to explain the location of heavy, raw material-based industry, especially when transport costs made up a large proportion of total costs. It still provides a useful aid to explain some location patterns, especially when later modifications of the model are considered. It is an essential part of some syllabuses, and so should be learnt, along with examples which illustrate how it works. However, later models, such as Smith's **maximum profit** model, are probably more useful in explaining present patterns.

After the industrial revolution industry was often tied to raw material and railway/port/canal locations. These are still important, especially in areas of heavy industry, and areas with poorly developed road systems. However, modern industry tends to be lighter, to use fewer raw materials, and to use road transport because of its speed and flexibility. Such industry is termed **footloose** and its location tends to be influenced by access to market, labour, or to government subsidy. New industry in EMDCs is mainly located near to major roads, on the fringes of built up areas. This often leads to **urban blight** in the old inner cities.

The capital for much modern industry is provided by **multinational companies**, which are able to transfer production quickly from one country, or continent, to another. In an EMDC the company is likely to locate where the financial inducements from the government are greatest. In an ELDC the location is likely to be the primate city, or the major port. In either case the company is likely to locate in a country where labour conditions are most suitable. This may be where the labour force is best educated, or it may be where the laws are most 'business friendly' or, in other words, give fewest rights to the workforce.

There are often benefits to be gained from the **agglomeration** of industry, or locating close to other firms. The **multiplier effect** then operates and provides economies for all the firms there. However, for some firms this may prove to be a disincentive. Competition for labour may push up wage rates, etc. and then they find benefits in **deglomeration**, or moving to a new, undeveloped area.

In some ELDCs the **informal sector** of the economy provides many important industrial jobs. This industry produces on a small scale, with low inputs of capital, often family labour, free from bureaucratic regulation. Much recycling of waste material takes place, sometimes using **intermediate technology**. It often meets people's basic needs, and serves a local market. It provides income for many people, and allows them to develop basic industrial skills.

Industry and the environment

The very nature of industry means that there is always a risk of damage to the environment. Manufacturing concentrates raw materials, and removes impurities as waste. The acts of concentration and waste production will be damaging, unless great care is taken. Even then, the long-term consequences may not be predictable. Some of the forms of damage are:

- ▶ **Derelict land** which has been used by industry, and then abandoned. Extractive industries are particularly damaging in this respect.
- ▶ **Waste heaps** which are produced by extraction and manufacturing. These are usually unsightly, but often contain noxious chemicals in high concentrations. The cost of disposal, often by burying in the original areas of extraction, is high.
- ▶ **Water pollution** by chemicals, or by waste heat from power stations.
- ▶ **Air pollution** by gases, or by noise. The former adds to the greenhouse effect, acid rain and photochemical smogs.
- ▶ **Industrial injuries** occur regularly from exposure to dust, noise, repetitive and heavy work, etc.

In the past the greatest pollution has been in what are now the EMDCs. However, they now have the technology, and can afford to limit industrial pollution. Much of the worst damage at present is caused by Newly Industrialising Countries (NICs), and by the former Communist countries of Eastern Europe. Many people feel that the EMDCs have a moral obligation to make their technology available free to the NICs and ELDCs, so as to reduce the pollution risks to those countries, and to the global environment.

Service industries

Service employment has traditionally been concentrated in urban centres. In most EMDCs it is by far the largest area of employment.

Services can be arranged in a hierarchy with **low order** services being needed regularly, and **high order** services being required less often. From this a hierarchy of service centres has developed. Low order centres (like villages and suburban shopping parades) have few functions, but occur very regularly throughout populated areas. They have small market areas, but the local people make frequent visits. The highest order centres (usually capital cities) have a wide range and number of services. People use them infrequently, but they come from a large area, and therefore in large enough numbers to support the services. Such services have to be concentrated in the most accessible places, in order to attract the necessary users from a wide enough area.

In recent years, in EMDCs, the use of shopping centres in CBDs has begun to break down. In the '**post-Fordist city**' the majority of people own, or have access to, cars which they use to obtain their needs. They no longer have to rely on public transport to the CBD. Instead they can travel to shopping malls, retail parks and superstores on the edge of town, where land is cheap enough to allow ample private parking, either free or very cheaply. This is leading to rapid changes in town structure, and is being matched by similar trends in entertainment and other services. The latest trend is for football clubs to move from the old working class areas near town centres to 'greenfield sites' on the edge of town, near to major road junctions for easy access by car.

The tourist industry

This is a popular area of study on many syllabuses. It is the world's second largest industry (after agriculture) in terms of money generated, and is the largest employer in many countries. Tourism relies on special resources, which are unlike those of any other industry. These include:

- sun
- sand
- beautiful scenery
- peace and quiet
- ancient monuments
- clean air
- historical links
- exotic culture
- novelty value
- good nightlife
- wild animals

Some of these resources are very perishable. In fact, in some cases, as soon as the resource starts to be used it starts to lose its value. Tourism is often seen as a good way of developing the economy of an area, but this can cause problems. Resort areas often go through a cycle of:

- **Discovery** by a few pioneers and explorers.
- **Exclusive development** when facilities are provided, at considerable expense, for the rich part of the market who demand exclusivity and novelty.
- **Mass development** as more and more facilities are provided; the price comes down, and the area appeals to a larger market.
- **Decline** when the area is ruined by its own success. It falls out of fashion, but as the number of visitors falls, attempts are made to attract a new market, with new specialised facilities, such as theme parks, or conference venues.

Unfortunately, such a cycle can leave large investments wasted, and many people unemployed. Even while the tourist industry is at its height there can be many problems:

- **Seasonal unemployment**, which may be eased by offering cheap deals to extend the season, bringing in workers for the peak season, etc.
- **Destruction of original economy** by using up farm land, attracting the workforce to new jobs which later leave them dissatisfied with traditional work, etc.
- **Social unrest** by bringing rich Westernised tourists into a poor, traditional area, with different customs. This can dilute or destroy the old culture.
- **Revenue leakage** as much of the money spent in the area finds its way back to the EMDC, through payment of profits, and payment for imports, etc.
- **Unfair employment practices** where the native workers can only get poorly paid, menial jobs, and many of the better jobs go to outsiders.
- **Environmental damage** to coral reefs, wildlife, wilderness, etc.

To counter such damaging developments **green tourism** is becoming fashionable. Here, a deliberate attempt is made to make the tourism as unobtrusive as possible. It should exist in a sustainable form, in harmony with local people and the environment.

REVISION ACTIVITY

This activity is designed to help you learn the factors influencing industrial location, and to prepare a series of examples which you will be ready to quote in the examination.

Copy and complete the table on p. 59. Give details of industries which have located in response to each of the factors listed. Wherever it is possible, use examples that you are familiar with from case studies you have learnt.

Note: answers have not been provided for this question.

Location factor		Name/ type of industry	Location	Explanation
Power supply:	Fossil fuel	●	●	●
	Alternative	●	●	●
Raw materials:	Mineral	●	●	●
	Agricultural	●	●	●
Labour force:	Cheap ELDC	●	●	●
	Cheap EMDC	●	●	●
	Skilled	●	●	●
Market:	Mass market	●	●	●
	Specialised	●	●	●
Environment		●	●	●
Transport:	Port	●	●	●
	Rail	●	●	●
	Road	●	●	●
	Airport	●	●	●
Government policy:	State investment	●	●	●
	Foreign investment	●	●	●
	Internal, private investment	●	●	●
	To tackle unemployment in declining region	●	●	●
	To spread development to periphery	●	●	●
Industrial inertia		●	●	●

PRACTICE QUESTIONS

Question 1

(a) Examine the methods by which agricultural output can be increased. (8)

(b) Discuss the economic and environmental consequences of increases in agricultural output. (12)

(London)

Question 2

Illustrate how the influence of any one factor, or group of factors, on the location and distribution of manufacturing industry may change over time. (20)

(WJEC)

Question 3

Study the table, overleaf, which shows employment in two regions of a developed country.

(a) (i) With the aid of information in the table, comment on the ways in which the employment structure in each region has changed between 1950 and 1990. [9 lines] (8)

(ii) Suggest reasons for the general shift to a higher percentage employment in tertiary industries in the country as a whole. [4 lines] (3)

(b) Why, in many developing countries, is the percentage employed in manufacturing relatively low? [4 lines] (3)

(c) Comment on the economic and environmental consequences of the tendency for manufacturing industry to be concentrated at the coast in many developing countries. [8 lines] (6)

(London)

Percentage of employment

| | 1950 | | | 1990 | | |
Sector of activity	Primary	Secondary	Tertiary	Primary	Secondary	Tertiary
Region A	55	30	15	10	45	45
Region B	80	15	5	50	15	35
Country	50	25	25	25	30	45

Location quotients

| | 1950 | | | 1990 | | |
Sector of activity	Primary	Secondary	Tertiary	Primary	Secondary	Tertiary
Region A	1.1	1.2	0.6	0.7	1.5	1.0
Region B	1.6	0.6	0.2	2.0	0.5	0.8

Question 4

Transnational corporations have brought about major changes to the organisation of business both within and between countries, and, using new technologies, to the way in which information, investment and products are transferred around the world.

Discuss the effects of these changes, and the attitudes of individuals and organisations to these effects.

In your answer you should include consideration of:

► the nature of the changes that transnational corporations have brought about;
► the sources of evidence used to identify the effects of these changes;
► the various effects of the changes and attitudes to them.   (25)
(*NEAB*)

Question 5

(a) Examine the impact of developments in communications on the tourist industry in recent decades.   (8)

(b) Study Figure 7.2 which shows factors governing the economic impact of tourism. To what extent can it be argued that tourism can stimulate economic growth in developing countries?   (7)

(c) Discuss the suggestion that tourism nearly always leads to the erosion of cultural identity. (10)
(*UCLES*)

Figure 7.1

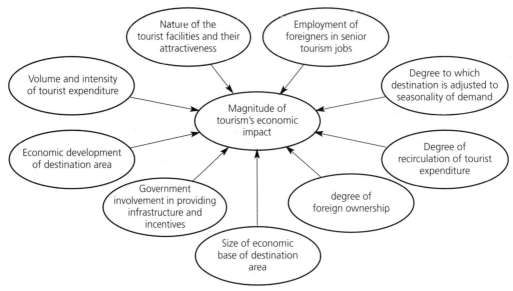

8 *World and regional development*

It is important to attempt to define economic development. It obviously involves growth of a country's **Gross National Product** (GNP). This is the value of all the goods and services produced in a country per person, per year. Figures for GNP range from over $US 20 000 (e.g. USA), to less than $US 150 (e.g. Tanzania). Measures of development closely linked to GNP can also includes factors like:

- energy consumption
- value of exports/head
- birth rate
- life expectancy
- food intake/head
- literacy
- % employed in manufacturing
- cars/1000 population
- death rate
- infant mortality
- doctors/1000 population

Many of the above measure 'quality of life' rather than pure economic development. Some people claim that improving the quality of life is more important than straight economic growth. In fact the UN produces a **Human Development Index** (HDI) which combines life expectancy, education and income, to give an index between 0 and 1. The HDI is closely dependent on GNP, but it gives a more rounded view of development.

Other attempts to measure development have included consideration of human rights and environmental quality, but these are very difficult to measure. In any consideration of development, though, it is important to ask the question: 'development for whom?' All the measures above give average figures for the population as a whole, but development should be about the poorest in society. The distribution of wealth is as important as the total GNP.

Paths to development

During most of the twentieth century the two main paths to development have been the **capitalist** model, where market forces are the main factors determining the priorities for growth and investment; and the **communist** model, where centralised planning of the economy is dominant. Despite the collapse of communism in Eastern Europe there are still countries which are attempting to achieve some degree of central planned development, especially in China, Cuba, Zimbabwe, some Indian states such as Kerala, and others.

Attempts by EMDCs to assist the development of ELDCs have, in the past, been based on the **modernisation approach** which involves industrialisation, development of an export sector of the economy, and other forms of capital investment. This has often led to little improvement in the quality of life for the majority. Now some organisations are adopting a **basic needs approach**. This refers to improvement of food, housing, health, water and sanitation for the world's poorest people.

Latest theories of development involve **ecodevelopment** or **sustainable development**. Only change which results in a stable habitat, where finite resources are not being used up, and where the environment is not being polluted, should be encouraged. Such an approach involves a very major shift in attitudes from most people and governments involved in development.

There have been some successes in the development process in the last 50 years but, unfortunately, it is clear that the gap between the richest and the poorest countries is still widening.

Theories of development

Rostow

This model suggested that countries could break out of the cycle of poverty by moving through a series of stages. The diagram below shows the Rostow model.

Figure 8.1
The Rostow model

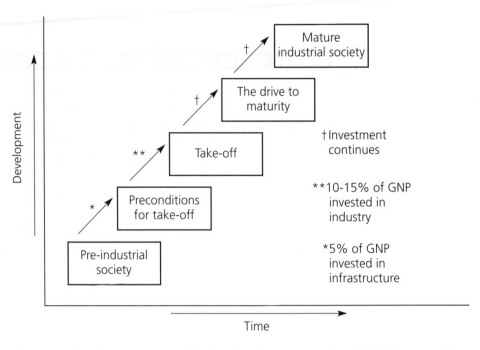

Movement from one stage to another involves investment of capital which comes from savings, or from outside. Investment of 5 per cent of GNP is needed to establish the infrastructure which gives the preconditions for take-off; and then the level of investment has to increase further for take-off, when one or two sectors of the economy become developed. The drive to maturity involves spreading development to all sectors.

When the countries of Western Europe and the USA developed there was capital available, either from exploitation of colonies or from the opening up of new resource areas. This is rarely the case in the ELDCs today. In fact population growth tends to use up any spare resources, so investment has to come from outside if take-off is to be achieved. This can cause huge debt, as in the case of Brazil and Mexico.

Core-periphery

Friedmann recognised that the world was divided into developed or **core regions** and **peripheral regions**. This idea can also be applied at other scales, with cores and peripheries within continents and countries. Development stems from imbalance. If resources are concentrated in the core then development is rapid. Eventually wealth will trickle down from the core to the periphery, from rich to poor countries, and from rich regions to the poor areas.

Myrdal took a more pessimistic view. He thought that the **spread effect** was more than counteracted by **backwash effects**. This means that resources, including the most educated and go-ahead people, are drawn from the poor periphery into the developing core. This leads to a cumulative process and a widening gap.

Frank went further. He said that core regions dominated the dependent peripheries. His **dependency theory** suggested that the process of development in one area *caused* underdevelopment in other areas.

Aid, trade, investment and development

Basically these are all ways of helping to make capital available for development. **Short-term** aid is usually designed to ease the effects of a disaster or to avert an imminent disaster. It usually takes the form of food, medical supplies, shelter, transport, etc. It keeps people alive, but makes little contribution to development.

Long-term aid is investment in real development. Long-term aid can be provided by:

▶ **Private agencies** Such as Oxfam and Save the Children Fund. It is usually on a small scale, and often designed to meet basic needs of the poorest groups in a country.

▶ **Governments** Either in the form of gifts or loans. Sometimes such aid has 'strings attached', such as insisting that the money be spent in the donor country, or being offered in return for purchases of arms. Such aid can support small scale projects, but is more often used on large, prestige developments, like dams and ports.

▶ **International agencies** Like the World Bank and the United Nations. Such aid often involves complex bureaucracy. The World Bank is especially noted for its insistence that receiver countries run their economies along strict monetarist lines. It often means there is little immediate benefit to the poor.

Investment can come from governments or firms. Such investments are expected to provide a profit for the investors. In some cases recipient countries have obtained great benefits. This has been particularly true in many of the economies of the Asian Pacific, where investment led to 'take-off' in the 1980s, followed by self-sustaining growth in the 1990s. These countries are known as the **tiger economies** or as **newly industrialising countries** (NICs).

In other cases the ELDCs have been exploited and have had their resources used up. They have seen large profits repatriated to the country of origin, and then the country has been abandoned by the investor company.

In the 1970s the **Brandt Commission** was set up to look at the problems of world development. They concluded that the biggest factor stopping development of 'The South' was the unfair conditions of trade. Governments and corporations in 'The North' were able to control trade in most commodities. They were able to buy raw materials from the ELDCs at low prices, because there were many countries competing to sell. If the prices of a commodity became too high they were often able to find a substitute. Meanwhile the price of manufactured goods rose steadily, and the EMDCs often put tariffs up to keep out imports of cheap manufactured goods, and to protect their own industries. The ELDCs fell into debt trying to buy the goods needed for development, and had to export more and more raw materials to try to keep up. Few were ever able to invest enough to approach take-off.

The Brandt Commission urged the EMDCs to improve the terms of trade for the ELDCs, and to invest in their economies. Since the report, however, there has been little evidence of improvement.

Some development strategies

Industrialisation on a large scale has been attempted by some countries. In the 1960s and 70s Brazil seemed to be achieving success, but the price they paid was huge debt, and they were forced to reduce their welfare payments and cut wages drastically because of the debt. South Korea, Singapore and Hong Kong have been more successful in the 1980s and 90s. They have achieved a high and sustained level of growth. In the early stages of development they were very low wage economies, with highly disciplined workforces. As the countries develop there may be pressure to raise wages and allow more freedom, which could cause development to falter.

Import substitution is when countries try to develop their economies by setting up factories to produce the basic needs that used to be imported. Food processing, clothing, construction materials (such as cement), and so on are the priority. This allows the development of a factory culture, and it is hoped that the early industry will be followed by more sophisticated, high technology industry.

Export-led growth tries to take advantage of the low wage rates in an ELDC to produce labour intensive goods for sale in EMDCs. The designer clothing industry and the electronics industry have been the two best examples of this trend. In the early stages equipment is mainly imported, but the industry may reach the take-off point when it can start to produce its own machinery. It may even move to the stage where it carries out the research and development functions that used to take place in the EMDC.

Agricultural development has included the Green Revolution with its intensification, based on high inputs. Mechanisation (or tractorisation) and the development of agricultural support industries were important aspects of the revolution. However, the green revolution led to a reduction of the agricultural workforce and widespread rural unemployment. It caused considerable social disruption.

In other areas **rural development** has meant land reform where large holdings, often owned by absentee landlords, have been divided between peasant farmers. **Cooperative development** of marketing has been encouraged, to give the small farmers greater bargaining power. Such development often uses intermediate technology, to meet people's basic needs. An essential part of the support needed for such small scale development is access to reliable, cheap loans.

Equivalent **urban development** to meet basic needs can be seen in the informal sector of many ELDCs. Small scale industries, with low capital, simple organisation, recycling of materials etc. can be seen in such developments as the Kula Jelai workshops in Kenya. If they are to make an important contribution to development they need access to money for investment, and the government has to encourage them to grow.

Regional development

Many of the ideas on international development can be applied on a smaller scale to regional development within EMDCs. It is quite possible to identify core regions and peripheries in the UK or in the EU as a whole, as shown below.

Core	*Periphery*
North and West Germany	Southern Italy
North and East France	Greece
Benelux and Denmark	Southern Spain
Southeast England	Portugal
Northern Italy	Former East Germany

If the core is not to go on developing at the expense of the periphery, investment has to be transferred to the peripheral regions. The EU Social Fund and the Regional Development Fund both do this. Within the UK money is made available through Urban Development Corporations, Enterprise Zones, hill farm subsidies, and other means of transfer.

Meanwhile free market economists argue that development will flow to the less successful regions if unemployment leads to low wage rates which will attract new industry. At the same time pressure to decentralise from the core will grow as the core becomes more congested and the quality of life falls.

REVISION ACTIVITY

Figure 8.2

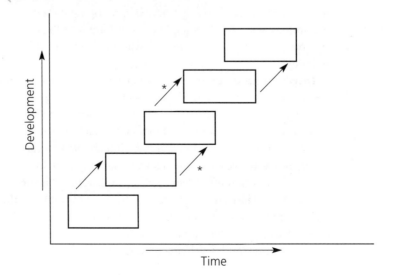

Figure 8.2 shows the framework of the Rostow development model. Draw it. Fill in each box. Explain what is needed at each point marked with an asterisk to move an economy from one stage to another.

Figure 8.3

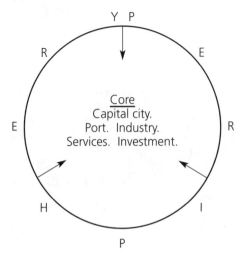

Copy Figure 8.3 of Myrdal's core-periphery model.

(a) Label the flows to show what moves from the periphery to the core.
(b) Develop another diagram(s) to show how the flow to the core could be reversed.

? PRACTICE QUESTIONS

Question 1
(a) Define the term 'newly industrialising country'.   (5)

(b) (i) Describe and explain the nature and location of manufacturing industries in one of the countries listed below:

| Brazil | India | Pakistan | South Korea |
| Taiwan | Philippines | Malaysia | Indonesia |

(15)

   (ii) Describe the impact of the growth of manufacturing industries on the employment structure of your chosen country.   (5)

(*Oxford*)

Question 2
For any *one* named country or large region in Europe, identify the characteristics which distinguish the core from the periphery and suggest reasons for the differences you have noted.   (25)

(*UCLES*)

Question 3
(a) State the evidence from Figure 8.4 on p. 66 that would support the following statements:

▶ there was a higher level of development in Indonesia than in Guinea Bissau in 1965
▶ there was a faster rate of development in Indonesia than in Guinea Bissau between 1965 and 1987   (4)

(b) Explain why it is important to use both social and economic criteria to measure development.   (10)

(c) With reference to specific examples, discuss the applicability of the demographic transition model to population change in developing countries since 1950.   (11)

(*Oxford*)

Figure 8.4

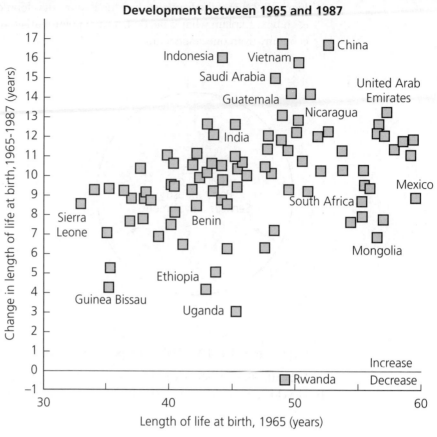

Development between 1965 and 1987

Question 4

Study Figure 8.5 which shows a model of manufacturing decline.

Figure 8.5

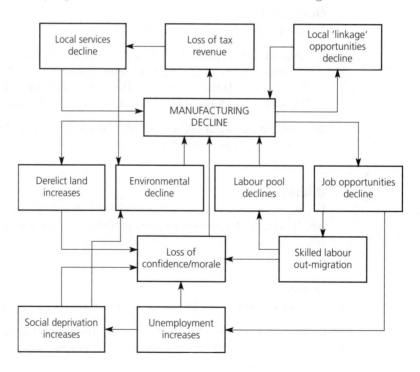

(a) For an area or region that you have studied where manufacturing has declined, assess the validity of this model. (10)

(b) For a newly industrialised country you have studied:
  (i) Outline the geographical factors which have promoted its economic growth;
  (ii) Examine the economic, social and environmental impacts arising from its economic growth. (15)

(London)

part III
All the answers

Solutions
Ecosystems

1

★ SOLUTIONS TO REVISION ACTIVITIES

1 Input a = minerals dissolved from atmosphere, in rainfall.
Input b = weathered rock.
Flow c = fallout as leaves and other tissue dies, excrement is returned to soil, etc.
Flow d = minerals taken up from soil, through plant roots.
Flow e = minerals released as litter decomposes.
Output f = litter carried away in suspension, or solution, by run-off water.
Output g = minerals leached out of the soil by water.

2 a = coniferous forest.
b = steppe.
c = rainforest.

ANSWERS TO PRACTICE QUESTIONS

Question 1 – Student's answer
There are eight separate parts to this question, and the part with most marks needs a diagram. Before you choose this question make sure you can do all, or most of it. You must not suddenly realise that you cannot do the last section – which happens to have the most marks. If you are sure you can tackle it well, plan your time sensibly. You have about 45 minutes for the question; there are 20 marks available; that means *roughly* two minutes per mark.

In part (a) of this question you are given only one line/letter. Short answers are needed, and a word or short phrase will be enough. Think clearly and sensibly. The first part of a structured question is usually quite straightforward.

(a) X   sunlight/rain/precipitation/atmospheric carbon
     Y   consumption/being eaten
     Z   death/decay/excrement ... etc.

(b) (i) Autotrophs are plants which convert the sun's energy into tissue. They do this by the process called photosynthesis. In this process carbon dioxide and water are combined to form carbohydrates, and oxygen is given off.

   Examiner's note Good. There are some clear definitions here. No words are wasted, and each sentence takes the idea a bit farther. Could a diagram have been used to make the answer clearer? The chemical equation is not essential in a geography examination, but it would have shown the process even more precisely.

   (ii) The size of each level in the system represents the number of individuals in that level. It is obvious that one plant of grass cannot support one zebra, and that a lion cannot survive all its life by eating one zebra. Therefore you need more individuals in the lower levels to support a few in the higher levels.

   Examiner's note This takes one idea and illustrates it quite nicely with a 'common sense' answer. It may be worth 2/4 marks. However, the answer below shows a much greater understanding of the processes of energy transfer. The idea was explained, and then two good illustrations were given. This was worth 4/4.

   (ii) There must be more individuals, and a greater biomass, in the lower trophic levels, because energy is lost at each stage of energy conversion. Hunting takes up energy, so there is a loss to the system. Not all the mass of a zebra is converted to lion tissue; the bones cannot be digested so energy is lost here too.

(c) (i) If one trophic level changes there will be knock-on effects for the other layers. If there was a drought there would be fewer autotrophs. This would mean there were too many herbivores, so some would die ... and so on up the levels.

Examiner's note This answer is a good attempt to develop a fairly complicated idea without running out of space, which is often limited in structured answers like this. But look again ... is the candidate actually starting to answer the next part here? You must be careful about this.

(ii) If less vegetation grows the herbivores might overgraze it at first, reducing it even more. Then there would be a sudden, catastrophic fall in the herbivore population. Eventually a smaller population would be established, in balance with the reduced vegetation.

Examiner's note There were 2 marks available for this section. The first two sentences of this answer clearly gain 1 mark. The last sentence develops a further point very clearly.

(d) When European settlers went to Australia they took some rabbits with them. When the rabbits escaped into the wild they had no natural predators, so their numbers expanded rapidly, and they consumed large amounts of grass. This reduced the amount available for native herbivores, and their numbers fell rapidly.

Figure A1.1

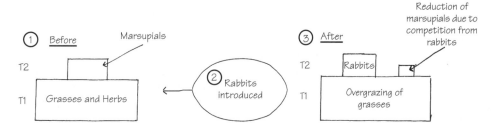

Examiner's note A basic answer would have described a simple change brought about by hunting, farming, etc. This answer goes farther than this, bringing in the idea of introduced species, and explaining their effects quite well. Unfortunately there does not seem to be much knowledge of the original species of plant, herbivores, or carnivores! The diagram is reasonable, but also a little vague. 4 marks were awarded.

Question 2 – Essay plan

At first glance this might seem like an example of a short essay title that does not provide any structure for the answer. At second reading it can clearly be seen to provide quite a lot of structure, or help with organising the answer. Use your highlighter pen, and break down the question.

Describe and *explain* the *natural processes of soil formation* that characteristically occur in *temperate environments*, and *discuss* how *human interference* might *modify* these processes.

The command words show that the question has three parts:

(a) (i) Describe
   (ii) Explain

(b) Discuss

Part (a) is about *natural processes of soil formation* , and (b) is about *human interference* and how it *modifies* the natural processes. The whole essay must refer to *temperate environments*. In other words, there is a fairly tight structure – but you are not told how many marks each part is worth, nor how much time to spend on each part. You must try to work things out carefully in the planning stage.

The following essay plan is more detailed than you could hope to do in an examination. However, you ought to practice writing such plans as part of your revision; then, once made, you can learn them and have them ready to write in the real examination. This plan sticks to the stages identified above.

(a) (i) provides a brief description, which does not go much beyond a list.
(a) (ii) gives a very detailed elaboration of all the points in the list, and then shows how these processes actually produce the main temperate soils.
(b) is briefer, but makes reference to how some farming activities can affect specific soil forming processes.

(a) (i) Describe the processes:

Weathering   • adds material at the base of the soil   [1]
                                                                                  (see notes below)

Vegetation growth and animal life
                                      • adds organic material at surface   [2]
                                      • carries material down into A horizon   [3]

Water moves through soil   • moves soil material with it, by:
                                         – mechanical downwash   [4]
                                         – leaching   [5]
                                         – podsolisation   [6]
                                      • gleying   [7]

Water can go up or down causing:
                                      • eluviated horizons   [8]
                                      • illuviated horizons   [9]
                                      • minerals leached away out of soil   [10]

(a) (ii) Explain the processes:

[1] In temperate climates there are few temperature extremes, so not much mechanical weathering. Freeze-thaw can be important on exposed rocks. Mainly affected by chemical weathering. Hydrolysis is the most important process, where water combines with rock minerals to form clay. This causes breakdown, even of hard rocks. Oxidation also important.

[2] Plants and animals provide organic material. This is mostly input at the surface. Forms: litter, then may form fermentation layer, and/or humus layer, then incorporated into A horizon.

[3] Carried down by water, or burrowing animals.

[4] Solid particles can be washed down through the soil.

[5] Leaching is when more soluble bases (N, Ca, Mg, Na, K) are dissolved and carried downwards by rainwater. Rate of leaching depends on acidity of water. More rapid under a humus layer. Minerals can be leached right out of the soil into rivers etc. [10].

[6] Podsolisation occurs in colder, wetter parts of the temperate climate regions. Poor organic decomposition leads to a very acid A horizon, so ground water becomes very acidic, and dissolves less soluble (Fe and Al) bases. This leaves a very bleached, grey layer.

[7] Gleying occurs in waterlogged soils. Iron salts are reduced. Brown soils become grey/black.

[8] Eluviated horizons have material washed out of them by leaching and podsolisation.

[9] Illuviated horizons have material deposited in them.

Soils formed by these processes in temperate regions include:

▶ **Brown earths**, formed under temperate, deciduous woodlands, with moderate rainfall and limited leaching.

continued

---

continued

► **Podsols** are formed on acid parent material, and beneath heathland coniferous forest.
► **Gleys** are formed where the soil is waterlogged. They can form on uplands, but also form on lowland marshy areas.
► **Peat** is formed where waterlogging is extreme. Deep organic layers form.
► **Chernozems and chestnut soils** are found under the temperate grasslands where the P:Pet ratio is low, or negative. Evaporation draws water towards the surface, leading to $CaCO_3$ deposition close to the surface. The grass produces a rich humus, and soil organisms carry the humus deep.

(b) Human interference:

Main form of interference is farming. Farmers produce outputs from their land, which always tends to remove some elements from soil. Ploughing mixes horizons, trying to bring new minerals to the surface to help crop growth. Addition of fertilisers and manure tries to replace minerals and organic material taken out by cultivation.

Draining of gleys allows air into the soil, in turn allowing organic material to develop. Allowing peat to drain can lead to development of very rich soil, as it makes large amounts of organic material available for plant growth. Interference with drainage can lead to unexpected leaching or gleying, depending on the nature of the interference.

Farming is not always beneficial to the soil. Over-farming can lead to loss of vegetation cover, followed by erosion of soil, at a faster rate than soil is replaced by weathering.

The length of this essay plan shows quite clearly that this could become a very long essay. How do you make sure that it does not take too long to write? Should you make fewer points in a lot of detail? Or should you cover as many points as possible, and maybe sacrifice some detail? You should make your decision based on the following advice:

► You should give reasonable time to each section, Describe, Explain and Discuss. 33 per cent/33 per cent/33 per cent is maybe not the right balance, because Explain is probably the most important. 25 per cent/50 per cent/25 per cent is the most you should tilt towards Explain.
► If you are aiming at an A or B grade you must make sure that at least some of your points are developed in detail. This is the only way that you can show high level knowledge and understanding.
► If you do not have a very secure understanding of soils, and are aiming to write just a 'reasonable' answer, you should probably try to make a series of simple, clear points. This is a reliable way of getting middle level marks, without wasting time.

Question 3 – Student's answer
Here you have to choose two contrasting seres. The main seres are found on:

shallow fresh water (e.g. lake margins)   hydroseres
shallow salt water (e.g. salt marshes)   haloseres
sand (e.g. dunes)   psammoseres
rock (e.g. volcanic eruptions, rock slips)   lithoseres

Any two of these would provide a contrast. Choose the two that you have learnt best. If you can refer to your own fieldwork on one of them, so much the better.

Ecosystems are communities of plants, animals and micro-organisms which exist within certain clearly defined climate and soil conditions. There are inputs into the system. In particular the sun provides an energy input, and heat is lost from the system as an output. Flows of energy move through the system, and nutrients are cycled through the system.

A sere is a succession of plant communities which develop and change over time. All seres are said to start on a newly exposed surface. At each of the stages in a seral progression the plants and animals cause changes to the soil, and these can lead to the soil becoming suitable for other species of plant, so a new community of plants becomes established. The new plants may replace the old ones, until a climax community becomes established.

A dynamic system is one that keeps changing. Obviously seres keep changing, so they must be dynamic. Even climax communities can change, if the climate changes, or if outside influences, like people, come in and start off sequences of changes.

The sand dune system that we studied at Studland in Dorset shows how seres change. I will contrast this with what happened on Surtsey after the volcanic eruption in about 1975.

Figure A1.2

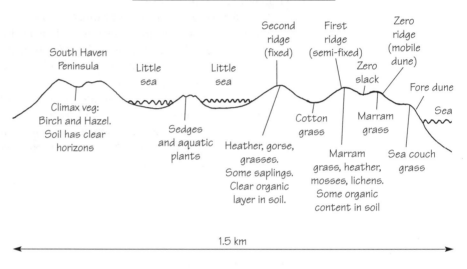

Studland Heath Dune – Vegetation Transect

1.5 km

At Studland the newest plant communities are found on newly colonised dunes near the sea. About 1 km from the sea are older dunes, with a climax community growing there. This climax community must once have been a newly colonised dune, and it must have gone through all the seral stages shown on the other dunes. Each seral stage created the conditions necessary for the next stage to develop. This has been a dynamic, evolving ecosystem. The newer dunes are still evolving, and are still dynamic.

> **Examiner's note** This candidate writes a very clear introduction where she defines the terms used in the essay, and justifies her choice of examples. Then she goes on to give a lot of relevant information in a diagram. She does not waste time repeating what is in the diagram, but makes some useful comments which refer to the diagram and link it to the essay.

Question 4

This question is largely skills-based. At least the first 11 marks are awarded for describing how you would prepare for and carry out a piece of fieldwork. If you can make reference to your own, real fieldwork you should be able to write a very good answer to the first two parts of the question.

However, you may have noticed that your examination board has a lot of this type of skills question; some boards have one on each section of the syllabus. Most schools do not have time to carry out full scale fieldwork exercises on all the topics; and so it might be worth preparing answers, based on how you could do field studies on those topics that you have not had time to actually study in the field. Some of the techniques, like the sampling in (a) here, are more or less the same, whatever topic you are investigating.

In part (a) of this question you need to show you are aware of the different ways of sampling:

▶ Random sampling. In this soil study you need to select a number of points, rather than selecting areas or lines, or picking from a list.
▶ Systematic sampling. This might be a quicker way of selecting. Will it give a fair sample?

► Stratified sampling. In this case you might think that soil would vary with height, so you might want to ensure that each height category was fairly represented.

Then you need to mention the size of sample needed. The larger the sample, the more likely it is to give an accurate representation of the parent population. But, on the other hand, you must be aware of practical constraints. The question does not offer much help with this. It does not say how much time is available, or how you should deal with problems of access. There may have to be a compromise between the ideal size of the sample, and practical constraints. However, this is a theoretical exercise, so I suggest you describe the best possible survey, and do not compromise too much! Note that this question asks about soil sampling. You could easily adapt this plan to fit sampling of vegetation, or land use, or land value etc. A bit more adaptation could make it suitable for choosing interviewees for a questionnaire, etc.

In part (b)(i) of this question there are 4 marks available. You have to write about two methods. That means that you have to score 2 marks for each method. You cannot be expected to write a lot for each method. Be brief! Be clear! Be concise! Here is an outline of the type of answer you are expected to write:

I have chosen to describe how I would measure soil texture and soil acidity. For both of these I would need to collect a sample in the field and take it back to the classroom for analysis.

To analyse the texture I would dry each of the samples in an oven. Then I would put the sample in a glass jar, add water and shake it so that all the soil was suspended. Then I would leave it to settle, so that the different sized particles formed separate layers at the bottom of the jar.

When the water was clear I would measure the thickness of the layers of sand, clay, silt and organic material.

I would have to be careful that my samples for testing acidity were not contaminated, by my hands or tools, while I was collecting them. I would put a small amount of the sample in a test tube, with distilled water. I would put a stopper in the tube (not my thumb) and shake vigorously. Then I would leave the sample to settle for a few minutes.

Then I would dip a piece of indicator paper in the sample, and wait for it to change colour. I would check the colour against the chart, and record the pH value of the soil.

Part (b) (ii) of this question has moved away from fieldwork now. It is testing knowledge of how soils vary over small areas. Of course you may be able to use knowledge gained in field studies, but that is not essential. It is useful to note that the 'stem' of part (b) gives quite a lot of clues about what you could include in this answer – 'texture', 'structure', 'acidity' and 'organic content'. The suggestion of looking at a valley transect is also useful. You are not 'commanded' to do that, but it is sensible advice, because most people studying soil will have considered how it varies down a slope. You can refer to acid leached soils near the watershed; thin soils on the valley sides; thicker, illuviated soils at the slope foot; and gleys on the flood plain near the river.

Part (c) is almost the same question as 1 (d) which has been discussed previously. Take the notes that were given for that question, and develop them to fit this question. Make sure that you include reference to the soil characteristics listed here in the stem to (b).

2 Solutions
River systems

SOLUTIONS TO REVISION ACTIVITIES

1 (a) 13 mm (b) 19 hours (c) 44 cumecs (d) 8.5 mm (e) 16 hours (f) 50 cumecs
(g) In the first storm the water fell on dry ground. Much infiltration, and
replenishment of storages occurs. In second storm ground already saturated; water
runs straight to river, so flood arrives faster, and reaches a larger volume. River still
coping with run-off from first storm.

2

Case 1 **B**   Reduced infiltration to bedrock storage, therefore increased
throughflow and run-off.

Case 2 **B**   Reduced infiltration to soil storage, and reduced throughflow,
therefore increased run-off.

Case 3 **A**   Faster run-off, therefore a lot of water arrives at river quickly, causing
rapid rise in level.

Case 4 **A**   Deforestation reduces interception and might reduce infiltration into
soil, therefore increasing run-off. Also removes root barriers to
throughflow allowing it to be faster. Removal of roots allows
increased soil erosion, blocking rivers, and reducing capacity.

Case 5 **B**   Many tributaries run into main stream. Time is not taken flowing
through a complex drainage system. Therefore a lot of water arrives
quickly in river.

Case 6 **A**   Surface is sealed, reducing infiltration and increasing run-off.
Surface is smoothed, reducing surface storage. Drainage system may
be put in place, further speeding flow of water to river.

Case 7 **A**   Sudden melt means large amount of water arrives in river, quickly.

Case 8 **B**   Furrows channel water straight down hill. There are no ridges to
impede flow, cause surface storage, and allow water time to infiltrate.
Faster run-off can increase soil erosion, leading to deposition in river
channel, and reduction in capacity.

Case 9 **A**   Peat is spongy and stores water. Disturbance of peat allows more
rapid drainage. Peat may be eroded after disturbance, further
reducing storage capacity. Peat is usually underlain by impermeable,
illuviated layer. This may be broken by ploughing, further increasing
the speed of drainage.

Case 10 **A**   Erosion reduces the storage capacity of the soil. Eroded material, if
deposited on river bed, reduces river's capacity.

ANSWERS TO PRACTICE QUESTIONS

Question 1
This title could hardly be simpler! However, you must still analyse it carefully, to see
exactly what you must do: 'With reference to your located case study (1), discuss (2) the
interrelationships (3) between factors influencing river discharge (4).'

(1) The syllabus for the London Board often makes reference to 'a located case study'.
As far as is possible the theoretical content of the syllabus has to be considered in
relation to a *single, specific* example. This is helpful advice from the board for teachers
and students alike. It narrows down the area of study, and makes it very clear what
needs to be revised. However, if you do not revise the example you have studied, you
will be severely handicapped in the examination. In other syllabuses students are left
free to study one or several examples. You must remember, though, that credit is always
given for references to real case studies, even when the question does not make that
quite as clear as this one does.

(2) You should be quite clear about the meaning of the command words (see the glossary on p. 9). 'Discuss' tells you to put forward different possibilities, and to write about interpretations of these possibilities. It needs careful thought.

(3) The idea of 'interrelationships' is absolutely fundamental to good geography. One factor affects others, and is affected by them in turn. You are looking for *links* between the factors, and for *causes* and *consequences*. In your answer these will be shown by use of connecting words and phrases, such as: '... and so ...'; '... therefore ...'; '... this leads to ...'; '... because of ...'; '... this results in ...'; '... eventually this produces ...'; '... in the short term the consequence is ...'; and so on. Every relevant use of such a phrase points to a relationship between two factors, and each such relationship raises the standard of the answer.

(4) The discharge is one of the main transfers in most river systems. Factors influencing this transfer include:

► inputs to the system
► the nature of the system and its transfers and storages
► other outputs from the system (mainly evapo-transpiration)

Once you have analysed the question you need to draw up a list of factors that you need to discuss. The list might look like this:

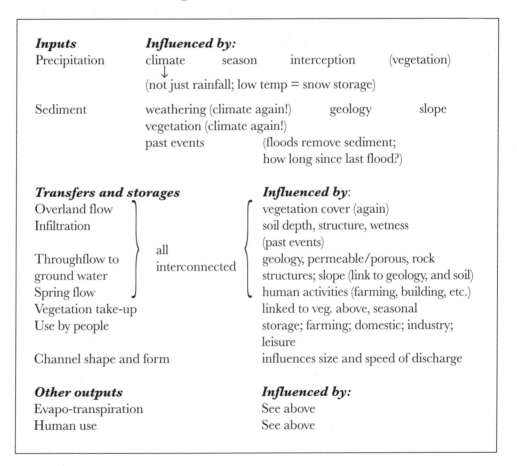

Inputs   **Influenced by:**
Precipitation   climate   season   interception   (vegetation)
                      ↓
                    (not just rainfall; low temp = snow storage)

Sediment   weathering (climate again!)   geology   slope
                    vegetation (climate again!)
                    past events   (floods remove sediment;
                                          how long since last flood?)

Transfers and storages   **Influenced by**:
Overland flow   vegetation cover (again)
Infiltration   soil depth, structure, wetness
                                     (past events)
                          all   geology, permeable/porous, rock
Throughflow to   interconnected structures; slope (link to geology, and soil)
ground water   human activities (farming, building, etc.)
Spring flow   linked to veg. above, seasonal
Vegetation take-up   storage; farming; domestic; industry;
Use by people   leisure

Channel shape and form   influences size and speed of discharge

Other outputs   **Influenced by:**
Evapo-transpiration   See above
Human use   See above

It should be clear from the list above that there are many links and interconnections that could be described and discussed. However you choose and organise your topics, keep reminding yourself that you are writing about *interrelationships* affecting *discharge*.

Finally, in any 'discuss' essay, the conclusion is vital. You need to round off your discussion. Here is an example of a concluding paragraph, which draws together several ideas about the River Hodder in Lancashire. Each of the points had been discussed in detail in earlier parts of the essay. This final paragraph emphasises the integrated nature of all river systems, and this one in particular.

From the discussion above it is quite clear that the Basin of the Hodder is a typical example of an integrated system, where the discharge of the river is affected by many interrelated factors. The sudden floods result from heavy rainfall being transferred very quickly to the river. Steep slopes with fairly thin soils, large areas of impermeable rocks, and land use practices that have reduced vegetation cover, all contribute to the floods by reducing storage and speeding up run-off. The influence of time is shown by the fact that the worst flood did not follow the heaviest rain storm. Instead it came after a period of wet weather that saturated the soil, followed by a storm that was only *fairly* heavy. The base flow of the river comes from water stored in the peat soils of the moors, and from the areas of permeable rock in the north of the basin. As long as burning does not destroy any more of the peat, and as long as intensification of sheep farming does not destroy more vegetation, the integrated system will continue to provide a steady discharge most of the time, and the occasional floods will not get any worse.

This shows what a concluding paragraph ought to do. The main essay broke the Hodder Basin down into its individual elements. The conclusion draws them all together again, and gives a neat summary in the last sentence.

Question 2

Here you are offered two questions on drainage basins, and you can choose to do one of them. How do you choose which one to do?

At first glance the two alternatives seem quite similar:
▶ They both have a 13/12 split between the parts
▶ They both refer to 'a named drainage basin' in (a)
▶ They both ask 'how you would investigate...' in part (b)

How are they different? A asks for one or more diagrams, and B does not. However, it is always useful to include diagrams where possible, so this should not influence your choice much. Strangely, A asks you to 'describe and explain' in both parts, and B only asks you to 'explain'; but as explaining is usually seen as the more difficult task, this should not have too much influence on your decision either.

The key difference lies in the following phrases:

A	*B*
'seasonal variations in river flow'	'an integrated physical system'
'downstream variations in discharge and velocity of river flow'	'relationships between slope form, soil and vegetation'

You should have noticed one word that is present in both parts of A and not present in either part of B. That word is 'river'! In fact, both times it is used as part of the phrase 'river flow'. Answers to A, part (a), must be focused on the river itself, and must describe variations in the volume and/or speed of flow. An ideal way to start to answer this question is by drawing a graph to show monthly patterns of discharge. Alternatively cross-sections, showing how the wetted perimeter varies from season to season, could be drawn. Then the answer has to move on to explain why these variations occur. Factors like seasonal rainfall and temperature variations will obviously be vitally important here. The best answers will also go on to consider the extent to which vegetation varies with the seasons; the influence of soil and rock type in storing water and evening out the flow; the effect of slope in speeding up or slowing down run-off, etc. In other words, they will end up as seeing the river as part of an integrated system.

Answers to question B have to start by considering the wide variety of elements that go to make up the integrated system. It might be useful to refer back to the notes on Question 1 now. They show how a question similar to this one was answered. Many of the ideas and examples given there could have been used in an answer to B. However, note that question A started by asking for a description of the river, and ended by asking for the river to be seen as part of an integrated drainage system. Question B deals with the drainage basin as a whole, and answers must refer to weather, soil, vegetation, geology, land use, slope and so on – but the best answers will also mention the river. It is the river that integrates and draws together all the other aspects of the drainage basin.

To sum up:

A	*B*
mainly considers the river, but	considers the whole drainage basin ...
... also considers how other factors affect the river.	...with the river as just one factor among many.

The two part (b)s are about investigations. As with many such questions you need to describe and explain how you would (i) sample; (ii) collect data; (iii) display data; and (iv) analyse the data and conclude.

	A	*B*
Sampling	The question is about seasonal variations. Therefore the discharge must be sampled at different times.	The question asks about relationships that vary from place to place. Therefore the sampling must pick points in a set area. Line transect often used in valleys.
Collection	Measure river cross-section (a diagram will explain this well). Measure speed of flow.	Measure slope angle. Dig a soil pit to study structure. A quadrat survey of vegetation.
	In either question you must describe methods clearly but briefly. Mention possible problems, and ways to avoid them.	
Display	Discharge = X-section × speed. Graph; discharge against time.	Transect diagram with slope angles. Soil sections drawn. Graphical representation of species.
Analysis	You would only have time to write a very brief comment on what you might expect your results to show.	

Question 3 – Student's answer

Sometimes examination candidates need to define the terms of the question, and explain how they are going to choose from within a broader field. This candidate's answer shows how he set about defining what the question meant, and how he was going to tackle it. Note how logically he develops the answer, moving carefully from point to point. He also writes in short paragraphs, emphasising his careful, methodical approach.

The hydrological cycle shows how water moves between the sea, the atmosphere and the land. In its simplest form there are four stages in the cycle. In reality the cycle is much more complex. There are many other storages and transfers in the system, especially in what is shown as the run-off stage. In this essay I will describe some of the complications of that part of the cycle, which is known as the drainage system.

Figure A2.1

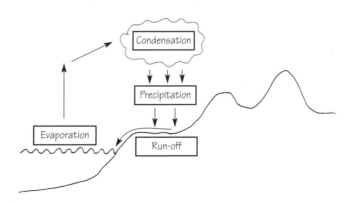

> **Examiner's note** The candidate knows what the whole hydrological cycle is. He names the four main stages, and also uses good technical language. His phrase 'many other storages and transfers in the system' shows that he is fitting this model into the general idea of systems structures. He then focuses in on the drainage basin system, explaining that the use of the term 'run-off' is not really enough to describe this part of the cycle adequately.

Precipitation can fall as rain, snow, hail, sleet, fog, mist, dew, frost, etc. When it reaches the land it can be stored, or it can move downwards under the force of gravity.

Some stores are natural, such as:

- ice caps, lakes on the land surface
- soil water and ground water stores below the surface
- storage in plants

Some are man-made, such as reservoirs, toilet cisterns and even bottles of beer.

All of these stores are temporary. Eventually the water finds its way out of the store, and becomes part of a flow again.

> **Examiner's note** This essay is being written in a very concise style, with little elaboration or development of ideas. However, it has a very logical structure. Information is being classified and presented very clearly. There is a rather light-hearted moment in the section on man-made stores. Should the writer have tried to be humorous? The examples are completely relevant, and the comment is not flippant, and so it is quite acceptable. Humour does not gain marks, so do not waste time over it; but used sensibly it may lighten the examiner's thoughts, so it need not always be avoided. Just be very careful.

The main natural transfers in the drainage basin are:

- run-off, which can be channelled in a river, or sheet flow across the surface
- soil throughflow
- rock throughflow

The proportion of water that travels by each of these routes depends on many factors, such as:

- the intensity of the rainfall
- the amount of vegetation cover
- how saturated the soil is before the rain starts
- how much space there is for recharge of the water table

The various proportions are important because they affect flooding. Run-off gets to the river most quickly. Soil throughflow takes longer to reach the river. Flow through the rocks usually takes longest (although flow through limestone caves can be very quick). So if a lot of the water flows over the surface there is likely to be a flood. When a lot infiltrates the soil there is less chance of a flood.

> **Examiner's note** The style is still concise and clear, but the candidate is making clear links and developing ideas. He gives brief explanations of the links, then he moves on to a new point. This is a very effective style.

Plants can take up water from the soil through their roots. They use the water in their growth processes. It passes up to the leaves, then waste water is passed out onto the surface. This is called transpiration. Water can be evaporated from the leaf surface, in a process called evapo-transpiration. Animals use water too, and waste water is lost as perspiration. This can also be evaporated, but it is not called evapo-perspiration.

The important point is that this is a kind of short circuit in the hydrological cycle. Water can be evapo-transpired, or evaporated from the surface of the ground, without ever going back to the sea. The diagram at the start of the essay is obviously over-simplified.

There are many ways that humans can interrupt the cycle too. Some have already been mentioned, and there are millions of other interruptions, but in the end the water either is evaporated (e.g. from power station cooling towers), or goes back into the rivers. It may be clean or badly polluted, but it will flow through the rivers to the sea (unless it gets evaporated), and complete the cycle.

Examiner's note This essay is a very good example of how to take a lot of fairly simple points and then use them to build a very good structure. There is no single, very profound idea, but the overall impression is of a candidate with a lot of knowledge and a clear understanding. The candidate's plan for the second part of the essay looked like this:

Raise the water table

increase inputs:
- pumping water into underground storage in areas of porous rock
- overuse of irrigation water e.g. USA in Colorado valley
- making reservoir like Aswan Dam, lets water infiltrate sandstone nearby

decrease outputs:
- reduce evaporation from surface by dry-farming techniques e.g. Negev Desert
- keep vegetation cover on farmland e.g. Amazon shifting cultivators use multi-cropping and inter-cropping to protect from evaporation

Lower the water table

decrease inputs:
- overuse of water elsewhere in the basin e.g. Mexico part of Colorado; e.g. Aral Sea
- over-grazing reduces infiltration, increases run-off. Water table falls, e.g. Sahel – can lead to desert'n

increase outputs:
- pumping irrigation water e.g. Deccan plateau in India – diagram of water table and well
- quarrying removes an impermeable rock and allows water to seep out, e.g. lignite mining in Ptolomaida, Greece, damages nearby farming

Examiner's note This plan provides a clear structure, again based on the systems approach. It is always helpful if you can base your planning on a model which is like a ready made plan, and save a lot of time in the examination. Each of the four parts of this plan has been developed, with two or three ideas and examples, usually clearly located.

3 Solutions
Plate tectonics

★ SOLUTIONS TO REVISION ACTIVITIES

1 constructive margin
- (a) oceanic plate.
- (b) mantle.
- (c) convection current.
- (d) old volcanic ridge, now reduced in height.
- (e) rift.
- (f) mid-ocean rift valley.
- (g) volcano on mid-ocean ridge.

2 destructive margin
- (a) volcano, forming offshore island, possible part of volcanic arc.
- (b) oceanic plate.
- (c) convection current.
- (d) mantle.
- (e) ocean trench.
- (f) earthquake foci, along friction area in subduction zone.
- (g) melting crust, in hot subduction zone.
- (h) magma reservoir, or batholith.
- (i) continental plate.
- (j) continental crust.
- (k) volcano.

3 conservative margin
- (a) fault line.
- (b) plate being dragged by movement of convection current in mantle.

ANSWERS TO PRACTICE QUESTIONS

Question 1

This is an extremely tightly structured question. It is divided into four parts, but note how many separate things candidates are asked to do in each part.

(a) describe *four* margins.

(b) suggest *three* pieces of evidence.

(c) *draw* a diagram, and *describe* the process*s* (plural).

(d) at least this only asks you to do one thing.

Comments are provided here followed by an example of a good answer for each part.

Space for part (a) is limited – and you *must* stick within the allocated lines (unless you are forced to cross out something you have written). Discipline yourself. Think very carefully before you start to write, and make every word count.

This is particularly difficult with margin A, because it lies very close to the point where the nature of the plate margin changes. D is difficult too. Some text books refer to this as a collision margin, but the question does not use this term.

Below shows how you could cope with these two problems:

A is where a **conservative margin** becomes a **destructive margin** ocean/ocean.

B is a **destructive margin** where an ocean plate meets a continental plate.

C is a **constructive margin**.

D is a **collision zone**, which is a type of **destructive margin**.

In part (b) you have to write about three pieces of evidence. There are two lines for each piece, and 2 marks available for each. So, in each small space you must make an elaborated point to gain full marks.

---

1 The outlines of the continents seem to match like a jigsaw. The fit is even better if the continental shelf is included.

2 Rocks of similar age, type and structure exist in SE Brazil and South Africa. They are so alike it is unlikely to be a chance happening.

3 Measurements of magnetism on the sea floor between S America and Africa suggest that the Atlantic has been spreading from the mid-ocean ridge.

Try not to repeat information from the diagram in your written answer. In fact, you should be able to gain full marks for part (c) by drawing a very good, well labelled diagram, even without any text. However, it would be unwise to risk this, unless you had great confidence in your diagram!

Note that the question asks about *processes*. It does not ask about features or landforms. Be sure that your answer is all relevant, and does not drift away from the question.

At destructive margins two plates are pushed together by convection currents. The heavier plate is forced downwards in a subduction zone. Friction between the two plates causes earthquakes, and generates heat which melts the plate material. The molten magma, under pressure, is forced back to the surface. It forms volcanoes. Sediments on the plates at either side of the margin are pushed up to make fold mountains.

You have six lines to answer part (d), and 6 marks to gain. On a more 'human' type of question like this it would be quite easy to waste space, by writing a 'chatty' answer in general terms. You must resist this temptation. Write about specific examples, and write in brief, clear but detailed phrases.

Mineral-rich soils are a main attraction of tectonically active areas. In Java very high densities of population are caused by the soils, which are ideal for rice growing. In Iceland and New Zealand people harness geothermal energy (as steam) to generate electricity, and heat homes and greenhouses. Plate movement can cause coastal indentations forming harbours like San Francisco. Tourism, the world's growth industry, can take advantage of spectacular scenery (Mt St Helens).

Question 2 – Student's answer

The first part of this question deals with the same subject area as Q1 (b) above, but note how different the approach is. Q1 asked for three suggestions; Q2 asks for an assessment of the evidence, and assessment is a high order skill. Refer back to the answer to Q1; then see how those points could be developed to answer Q2. This is one candidate's attempt, with comments from the examiner:

Point 1: Ever since the earliest world maps were drawn people have noticed that the continents seemed to fit together very well. Some people wondered whether the continents had once been joined, but the suggestion was not taken seriously. How could it be when people still believed that what was written in Genesis was the literal truth. Even when the outlines of the continental shelves were mapped, and Darwin started to develop the theory of evolution, there was no way of explaining how the continents could have moved, so the theory was dismissed.

Point 2: In the 1910s and 20s Wegener brought together information from a lot of disciplines. He found that fossil reptiles from South America had also been found in South Africa. It seemed impossible that such creatures could have swum across the Atlantic. There were also folded rocks on both sides of the ocean, which have very similar structures. They looked as though they might have been joined at one time. There was also evidence of glaciation on both continents – and that could not have happened if the continents had been in the same climate zones as they are now. Other scientists did not accept these views. They did not take Wegener seriously, because he was not a specialist. Also there was still no way of explaining how continents had moved.

Point 3: In the 1950s and 60s very detailed studies of the bed of the Atlantic were done. The Mid-Atlantic Ridge was discovered with very new rocks, and a volcanic zone in the centre of the ridge. Drillings of sediments on the sea bed were studied, and they showed that the magnetic pole had changed many times while the sediments were being formed. The sediments were in matching strips on either side of the ridge. This seemed to prove that the Atlantic had been widening as the sediments were being formed.

Carbon dating of the volcanic rocks below the sediments proved that the Atlantic had been spreading. It also showed how the movement could have been caused. It seemed like the ridge had been forced apart by volcanic eruptions, so this was the mechanism to move America away from Africa and Eurasia. Later studies showed that as the Atlantic spread there was compensation by destruction of plate material in subduction zones. Recording of earthquake waves also showed that there were fluid layers below the earth's crust. Convection currents could be travelling through these layers.

Conclusion and assessment: All the pieces of evidence found during this period, and since, fit together to support the theory of plate tectonics. No one piece of evidence was enough on its own, but when evidence from lots of studies all mounts up like this it is very convincing. All the geophysicists who assess the evidence seem to agree that plate tectonics theory shows that Africa and South America were once joined.

> **Examiner's note** If the question says that you must assess, then your answer must make a conclusion about the evidence. Very few A-level candidates will be in a position to carry out their own tests to assess the evidence for plate movements in the field. Here the candidate assesses how other people view the evidence. It is quite true to say that the theory is accepted by 'scientific opinion', so this answer makes a well-informed assessment, and rounds off the essay well.

Question 3

There are quite a lot of similarities between this question and the previous ones. In fact questions on plate tectonics may be slightly more predictable than those on some other topics. The topic means that candidates must understand certain ideas very precisely, but those ideas then underlie all the work on the subject, and are likely to be tested regularly.

(a) (i) ────── = Constructive margin
        ══════ = Conservative margin
        ▬ ▬ ▬ ▬ ▬ = Destructive margin

(ii) The West Indies is the obvious answer.

(b) In Q1 you had to draw a diagram of a destructive margin. Here you are asked for a constructive one; in Q1 you were allowed to supplement your diagram with text, but here marks are awarded for the diagram alone. Make sure that you include crust and mantle; convection currents pulling the two sections of crust apart; release of pressure allowing the mantle rocks to melt and flow to the surface; eruption of submarine volcanoes; growth of the mid-ocean ridge; and the spreading of the ocean floor. (That last mentioned point is often forgotten, but it is the key to gaining full marks.)

(c) (i) Almost exactly the same question was part of Q1.
    (ii) Here is an outline answer:

Area B – California. It is a highly developed area. People are ready for earthquakes, so few lives are lost. C is Azerbaijan, where buildings are poorer and they are less prepared. Death tolls are high, but less property is lost – there is less there to start with.

The difference between the effect of hazards on economically more developed areas and economically less developed areas is another popular theme for examiners. Be ready for it, with good examples to quote.

Question 4 – Student's answer

This question is not laid out in the structured way that many others are, so you may find it useful to structure it yourself, like this:
With reference to acidic volcanoes (for example):

(a) explain their possible causes
(b) explain their likely effects on:
   (i) landscape
   (ii) people

20 marks are available. You have no guidance on how the examiner will distribute them. It seems to me that explaining the causes is a test of understanding, and is probably the most important part. I suggest dividing your time in the following ratio: (a) 8 (b)(i) 6 (b)(ii) 6. 10:5:5 might be even better. Here is an example of a very well structured, and detailed answer given by a student:

Almost all volcanoes are found at plate margins. Convection currents in the mantle push the earth's plates sideways over the mantle. When pressure on the mantle is released the rocks become fluid and flow upwards as magma. This sometimes escapes at the surface to form extrusive volcanoes. When plates are pulled apart it forms a constructive margin, but the lava here is basic, and I intend to write about acid volcanoes in detail.

> **Examiner's note** This first paragraph contains a brief summary of the basis of plate tectonics. Each of the first four sentences contains a separate, general point. The fifth sentence focuses the essay on one type of volcano.

Acid volcanoes are usually found at destructive margins. Here two plates are forced together, and the heavier of the plates is subducted beneath the other one. As it sinks down there is friction between the two plates, and this produces heat, which melts the plate that is sinking down into the hot mantle. The pressure caused by the plate being pushed down into the mantle forces the molten magma back towards the surface.

Examiner's note Here is another paragraph with a number of short sentences. The first sentence makes a nice link back to the introductory paragraph. Then each sentence, and even each phrase of a sentence, makes a new point. The causes of the volcanoes are explained. Now the answer needs to explain why they are acid.

The material which comes out of these volcanoes is known as 'sial' — because it is made up, mainly, of silica and aluminium. It is derived from melted continental crust material. This contrasts with the basic 'sima' found at constructive margins, which comes from denser layers of the mantle and is made of silica and magnesium. The acid lava has a high melting point, so, as it cools, it turns solid quickly. Therefore it is slow flowing and viscous.

Examiner's note This paragraph ties together what has gone before, and provides a nice link into what we have called part (b).

The viscous lava turns solid quickly when it is exposed to the air. As the lava cannot flow far from the vent it produces a steep sided cone. The sides of the cones are often convex. Sometimes the lava turns solid so quickly that it actually blocks the vent. This can trap hot lava below the 'plug', and pressure builds up as it tries to escape. The worst volcanic eruption ever recorded was at Krakatoa in the East Indies, when the vent was blocked and the whole island was shattered by the eruption which followed.

Mount St Helens is another acidic volcano. The pressure here built up until a side vent formed. The rising gases caused a bulge on the side of the mountain, which slipped away in a landslide, allowing all the accumulated gas and ash to flow sideways out of the volcano. This was unusual. It is more common for alternate eruptions of lava and ash to form layers on the sides of the mountain. Mount Etna is typical of this type of volcano.

Such volcanic activity has an effect on the landscape long after the volcano ceases to be active. In the Massif Central of France there are many tall, steep sided 'puys' which represent the plugs formed in the vents of acidic volcanoes. The softer rocks from around have been eroded, leaving the hard plugs sticking up, dominating the landscape.

Examiner's note This section starts with general points about the shape of acid volcanoes, and then goes on to mention more specific detail. Reference is made to four different examples, each of which illustrates a different landscape feature.

On the top of some of the puys are churches and fortified buildings, which shows how people have taken advantage of what is a basically hostile feature. The damage and destruction that can be done by acid volcanoes is obvious. Even the best scientific monitoring cannot accurately predict when an eruption is about to happen. Death and destruction can be caused by the explosion, by the lava or ash, by the nuée ardente (hot gas flow) as happened at Mont Pelée, or by subsequent landslides and mud flows as with the eruption of Nevado del Ruiz in Colombia. This eruption melted the snow on the mountain. The meltwater mixed with ash from this and previous eruptions, and formed a mud flow which drowned 25 000 people in a nearby town.

Clearly acid volcanoes present a hazard, but the hazard only becomes a disaster when a large number of people live nearby; so why do people live in areas which are known to be so hazardous? Partly this is a result of population growth pushing people into ever more marginal, and dangerous, areas. But the volcanoes also provide some very positive attractions.

Many people live near volcanoes because of the mineral-rich soil for agriculture. The olive groves and intensive smallholdings on the slopes of Mount Etna are evidence of this, and so are the very high densities of subsistence rice farmers on the terraced slopes of Java. Volcanoes also supply minerals. Sulphur is the most obvious one, but the trade in permeable rocks for water filters around the extinct volcanic puys of Auvergne should also be noted, as should the sale of Volvic mineral water from that area.

However, the main contribution that volcanoes make to the economy of many areas is as an attraction for the tourist industry. The 'Parc des Volcans' brings many people to the Massif Central. An old mine on the edge of Puy Chopine has been turned into a geological and ecological trail for discerning visitors. Some less discerning ones spend a lot of money buying expensive, tasteless souvenirs – made from volcanic ash and lava!

The tourist industry around Mt St Helens has become even larger and more commercialised than the one in France. The fact that the volcano erupted fairly recently gives an added excitement to the industry, but the technology that is available means that any future eruptions can be predicted with reasonable accuracy, and the area could be evacuated safely if another eruption threatened – or so everyone hopes. Actually another benefit from this volcano has been the growth of the research industry. Many geologists are employed in attempting to predict future eruptions. These scientists bring further money into the area, to the benefit of the local people and economy.

> **Examiner's note** This last section could be cut shorter if time pressure got too much – but it deals with the damage done by acid volcanoes – again quoting a number of different examples – and then goes on to mention benefits. The references to the Massif Central are not found in any of the standard textbooks. This candidate seems to be quoting from her own experience, and this always gains credit, when relevant.

4 Solutions
Meteorology and climate

SOLUTIONS TO REVISION ACTIVITIES

1 an increase in height through the atmosphere; 6.5.
2 rise; expand; cool; sink; compressed; warm.
3 unsaturated, rising air cools; 9.8 (or 10).
4 latent heat; water vapour condenses; less; cumulonimbus; rain.
5 more quickly; dew point; cannot form; passing over high land, or meeting colder air.
6 less quickly; dew point/saturation; more slowly; S; latent heat.
7 Conditional instability; DALR; saturated; SALR.

ANSWERS TO PRACTICE QUESTIONS

Question 1

Part (a) asks candidates to compare statistics, and then explain the differences. Here are four possible answers, which compare the temperature statistics:

---

a Plymouth is warmer than Oxford in winter; Oxford is warmer than Plymouth in summer.
b Plymouth is warmer than Oxford in winter. In January its temperature is 7°C, compared with 4°C at Oxford. In August its temperature is 15°C compared with Oxford's 17°C.
c Temperature differences between Plymouth and Oxford are listed below:

J	F	M	A	M	J	J	A	S	O	N	D
3	3	1	=	=	=	−1	−2	=	=	2	3

For five months, from November to March, Plymouth is warmer; but in the two summer months of July and August Oxford is warmer.
d Plymouth's temperature range is 10°C, from 7°C in January to 17°C in July. Oxford has colder winters and warmer summers and its temperature range is 14°C – from 4°C in January to 18°C in July. Oxford has the more extreme temperatures.

Now compare the answers above:

Answer **a** identifies the key idea, but the statistics have not been used at all. Some credit would be given for this answer, but it could be much better.

Answer **b** quotes some statistics, and this answer would gain more marks than **a**. The figures used are the most important ones – but they are just *lifted* from the table. They have not been *manipulated* at all.

Answer **c** is very thorough. The statistics have been *manipulated* here, and they are well presented, so that a clear pattern emerges. This is a good, sensible answer, but is it selective enough?

Answer **d** uses the figures and is very selective. The concept of 'range of temperature' sums up all the information that **c** gives, and shows very clear understanding, resulting in an ideal position to go on to explain the reasons for the differences between the two sets of temperature data.

The explanation for the differences must deal with marine influences on Plymouth's temperature, contrasted with the effect of Oxford's inland position. The influence of the prevailing winds carrying the marine influence may have a slight effect on the two places, as Plymouth is in the southwest and receives the wind directly off the sea. The difference in latitude may be mentioned – but it is a very minor factor. Details of altitude are not given. It can be assumed that Plymouth is at sea level, and might be slightly warmer because of that – but this is also a very minor factor.

The answers **a**, **b**, **c** and **d** might have started to describe the rainfall difference like this:

> **a** Plymouth has more rainfall than Oxford.
> **b** Plymouth has 291 mm more rainfall than Oxford.
> **c** Oxford only has 70 per cent of Plymouth's rainfall.
> **d** Oxford has only 70 per cent of Plymouth's rainfall, but there is a much bigger difference between the two in winter than in summer.

They should all then have gone on to describe the seasonal distribution of rainfall:

> Depressions and westerly, onshore winds account for Plymouth having more rainfall than Oxford in all seasons. They also explain why there is a marked winter maximum. The depressions and westerlies also bring much of Oxford's rain, but the total is lower because of the inland position. The comparatively high summer rainfall in Oxford can be explained by the higher temperatures causing convective rainfall.

Part (b) of this section is still about the contrasts between coastal and inland climates. What else *could* be included? Few candidates in the examination will have studied either Plymouth or Oxford – so rapid thought and application of general principles is essential here. Below is a plan for an answer:

	Coast	*Inland*
Cloud	More – wind off sea – moist air forced to rise.	May be rain shadow – less cloud. Summer convective cumulonimbus.
Humidity	Usually fairly high – maritime air streams.	Likely to be lower.
Fog/mist	Advection quite common, esp. when warm SW wind blows over cooler sea.	Radiation possible, during winter anticyclones.
Ice/snow	Rare.	Common in winter – more extreme temps.
Wind	Exposed to wind. SW gales may be common.	More sheltered, so winds probably less strong.

For part (c) here is a warning – geography is about interrelationships. It is very dangerous to leave out sections of the course when you are revising. On what is essentially a climate question, 28 per cent of the marks go on vegetation, and its adaptations! It should also be noted that the question asks about adaptation to *climate*. This is not a question about sandy soils or tolerance of salt.

Note the climate characteristics of coastal regions referred to on the table above – moderate temperatures; cool summers; generally high humidity (although some coastal areas are deserts); mist and fog; frequent strong winds. Then explain how plants adapt to these, always remembering to include 'references to specific examples'.

Question 2 – Student's answer

In planning an answer to this (and many other questions) it may be useful to ask yourself these questions – What? Where? When? How? Why? Here is an essay plan from a candidate who did just that:

	Depressions	Cyclones
What?	Low pressure systems. Tm air pushing north into Pm and being lifted over cold air.	Intense low pressure systems. Air circles into centre of convection uplift.
Where?	N Atlantic – between 40° and 70°N. Front moves N in summer and S in winter. Move from W to E, across Europe. Predictable paths.	Tropical oceans – between 5° and 20°N or S. Give names. Move from E to W and away from equator. Erratic paths.
When?	Throughout year, esp. autumn and spring.	Mainly late summer and autumn, when sea is hottest.
How?	Low pressure at centre. Long period of moderate rain at warm front. Short, heavier rain at cold front. Winds blow anticlockwise. Variable strength, depends on pressure gradient.	Very low pressure at centre (ex. eye). Rain becomes heavier with approach. Torrential, then sudden stop in eye; sequence then reversed. Winds anticlockwise. Moderate on edge – v. v. strong at centre.
Why?	Waves form on polar front, probably caused by jet stream. Becomes warm sector. Cold front moves faster than warm front and lifts warm air off ground.	Sea >26 °C. Heats air. Much evaporation. Air rises at centre. More air drawn in, etc. Condensation releases latent heat. Earth's rotation increases the spiral effect of winds.

Examiner's note This plan allows both systems to be dealt with equally thoroughly. It allows the essay to develop both description and explanation – or, in syllabus terms, to show knowledge and understanding. The 'Where?' section allows the systems to be located clearly.

Such a 'key question' approach can be adapted as a way of structuring many rather open-ended or unstructured essay questions. The questions provide a template which can be altered slightly to fit many questions.

Question 3

The concepts of adiabatic lapse rates and air stability are difficult ones to understand. Many A-level students find this the single most puzzling aspect of the whole course. This question shows, though, that it is worth grappling with. Once the ideas have been understood then the questions on it are quite straightforward and predictable. Given an initial understanding, many candidates could pick up high marks on this question – but without clear understanding there is no way that an answer can be improvised here.

The brief answers given below are all adequate to show a *reasonable* understanding of the concepts involved. They may be brief, but they are concise:

(a) (i)   the DALR is the faster rate of cooling.

    (ii)   the SALR is the slower rate, when latent heat is being released.

    (iii)   the CL is the base of the clouds, when DALR changes to SALR.

(b) The ELR describes how air temperature changes with height in the particular air being described. It varies from one place to another.

(c) Both rates show how air cools as it rises, but the SALR is slower, because latent heat is released as water vapour condenses.

(d) (i)   Instability usually occurs on hot days, when air at ground level becomes very hot. It rises quickly producing towering cumulonimbus clouds. These can bring hail, heavy rain, thunder and lightning. Conditional instability only produces cloud and rain when hills cause additional uplift.

    (ii)   Stable air is often associated with high pressure, descending air, and anticyclones. This is of benefit to the tourist industry because:
1 – there is little or no cloud, and so plenty of sunshine, which is essential for many people's holidays.
2 – there is little or no rainfall, which would ruin people's holidays.

Question 4

Part (a) is quite a straightforward question. A definition of the energy budget, and an explanation of the main flows in the system is given on p. 30.

Parts (b) and (c) both require specialised knowledge of the appropriate climate regions in order to write full answers. However, some of the techniques needed for these questions are valuable for people who have not studied these particular regions in depth.

In part (b) foolish candidates will look at the graphs as new pieces of quite unfamiliar data. They will spot a feature on one graph, and then try to find a similar feature on another. They will work hard, may work out the patterns, but will spend too long puzzling over the graphs. The most foolish may forget that summer in the S hemisphere is during Nov–Feb! Wise candidates, on the other hand, will immediately refer to their bank of geographical knowledge where they already have a structure for this answer. As rainfall in tropical Africa is dominated by the ITCZ and its seasonal movement, wise people will look at the graphs to try to spot:

1 Equatorial climate – high total – rain at all seasons – maxima at equinoxes.
2 Tropical – moderate total – summer rain – dry season of 3–4 months.
3 Tropical margins – low total – summer rain – long dry season.

When candidates know what they are looking for, it becomes easy to divide the graphs into three groups: 1 = Libreville; 2 = Bathurst, Wagaduga, Luanda, Bulawayo and Alavi; 3 = Timbuktu and Khartoum. Mogador does not fit the pattern. It is on the west coast, 30°N, and has an almost Mediterranean-type climate. Once the pattern has been worked out logically, using geographical understanding, then the justification of the decision is fairly easy.

Part (c) appears to be a difficult question. This is a 'compare' question in some respects, because candidates must assess similarities and/or differences between monsoon and other climate types. The examiners expected candidates to know that some parts of the monsoon region are very similar to some parts of the equatorial and the tropical climate regions, but that there are significant differences in other parts. Fortunately, though, the question does not demand an explanation of the causes of the climates. The mark scheme recognises three different levels of pass mark, and it is instructive to see exactly what was needed to achieve each level.

Level 1 shows only a very general awareness of monsoon climates (e.g. no appreciation of differences within the region). Other tropical climates also seen in very simple terms, but at least the main temperature and rainfall characteristics identified. Limited argument with no development of reasons.   (4–5 marks)

Level 2 distinguishes variations between monsoon and other similar tropical climates in such features as precipitation amounts and regional factors. Uses knowledge to justify any line taken on the question.   (6–7 marks)

Level 3 additionally points out that some tropical areas outside the classic monsoon area can have similar characteristics (e.g. Madagascar) but some parts of SE Asia have climatic conditions untypical of the tropical norms.   (8–10 marks)

Question 5 – Student's answer

All good geographers dread having to listen to 'saloon bar chat' about a geographical topic that is in the news, that all members of the general public have an opinion on, and about which very few of them have any detailed knowledge. These people do not let their lack of knowledge stop them from expressing their opinions very forcibly to anyone who will listen. On the contrary, they often refuse to listen to people who can talk with some knowledge on the topic. Maybe they fear that facts would ruin a good story.

Global warming is one of those topics, and some candidates write essays that are very like saloon bar chat. Examiners often feel that such candidates have picked up their scraps of information and opinion from sensational press coverage, and not from careful study of balanced views. You should be advised to avoid topics which you have not actually studied, but which you vaguely remember something about, from a TV programme that you half watched a few months ago. Such answers are usually awful! Here is an example of a good answer to this question, written with sound knowledge:

Since the 1960s scientists have been building up more and more evidence that global warming is taking place. Although it is still not absolutely certain that it is happening, most informed people now agree that it is probably taking place, and that human activity is probably a major cause of the warming. Over the earth as a whole there was a general, slight rise in temperature between 1860 and 1960, but this was within the limits of temperature change that occurs naturally. Since 1960 the warming seems to have got quicker, and nine out of the ten hottest recorded years have come since 1964.

> **Examiner's note** This is a good introduction which puts the problem in context. It avoids the sensational approach, but provides a useful set of figures.

Global warming is probably caused by the 'greenhouse effect'. As the diagram shows, changes in the composition of the atmosphere have caused an increase in the amount of energy trapped in the atmosphere.

Figure A4.1
The greenhouse effect

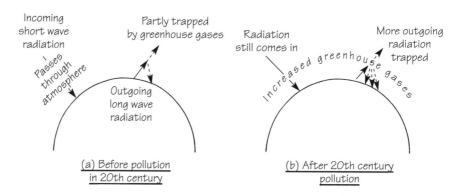

(a) Before pollution in 20th century   (b) After 20th century pollution

> ***Examiner's note*** Here a diagram is used to show a lot of information clearly. It is always good to learn such diagrams. They help structure learning, and make recall easier. Then the candidate avoids the common fault of repeating in text what has been shown on the diagram. This would have wasted precious time.

Some of the greenhouse gases occur naturally. Carbon dioxide is circulated between plants, the earth, the sea and the atmosphere in the carbon cycle, and methane is produced by decaying vegetation. However, human activities have speeded up the production of both gases. Burning fossil fuels has made the biggest contribution, and has been going on – at an accelerating rate – since the industrial revolution. Burning forests has also been an increasing problem recently. Both these processes release carbon dioxide much faster than natural processes would do, and so disrupt the stability of the cycle.

The other main greenhouse gases are almost entirely man-made. Nitrous oxide comes from the manufacture and use of fertilisers and other chemical products, and also from car exhausts and power station chimneys. CFCs are made for aerosols, fridges, solvents and foam rubber, and although they are being phased out some people think that their replacement chemicals also damage the atmosphere. It is probably true that the EMDCs have done better with phasing out dangerous chemicals than the ELDCs have, because they can afford the technology more easily. However, we cannot feel smug and blame the poorer countries for the problem. If it is to be tackled then the richer countries must make their technology available freely to all countries. The poor will stay poor, for the sake of protecting the environment that only the rich can afford to enjoy.

> ***Examiner's note*** There is good detail of four causes of the problem here. The distinction made between more and less developed countries is a good geographical point.

There seems to be agreement about the facts of global warming, and probably also about its causes, but meteorology is so complex that even the best computer simulations cannot produce agreement on its probable climatic consequences. It is probable that the increased heating will cause the seawater to expand, and this will lead to a rise in sea level.

Some people think that warming will also melt the polar ice caps; but others say that the increased temperature will lead to more evaporation, so greater snowfall, which will counter the effect of the melting. Both sides of this argument point to glaciers in their support, with one side saying that Alpine glaciers have retreated by hundreds of metres this century, and the other side saying that glaciers in Alaska are advancing quite rapidly at present.

Rainfall will also be affected. Changes in temperature may lead to a general migration of the world's pressure and wind belts. It may be that some areas will receive more rainfall, but some predictions say there will be a considerable reduction in rainfall at about 40°N – which will have a damaging effect on some of the great grain producing areas.

Another common prediction is that the increase in energy in the world's circulation system will lead to increased frequency of dramatic and extreme events, like hurricanes and other storms. There are frequent articles in the media claiming that each new weather event is a result of global warming, but these are probably over-dramatisations – so far, at least.

> ***Examiner's note*** Several possible changes are discussed, but nothing is overstated. The discussion is well balanced, and the essay ends with a nice, brief touch of cynicism, which left this examiner smiling!

5 Solutions
Population and resources

SOLUTIONS TO REVISION ACTIVITIES

1 Answers are given within the text in Chapter 5.

2

	Stage I	*Stage II*	*Stage III*	*Stage IV*
Birth rate	high	high	falling (S curve)	low
Death rate	high	falling at an increasing rate	falling at a decreasing rate	low
Population change	stable (or fluctuating)	rising at an increasing rate	rising at a decreasing rate	stable
Total population	low			high

(a) (i)   death rate starts to fall
    (ii) birth rate starts to fall
    (iii) population stabilises

(b) (i)   improved medicine, hygiene, food distribution, housing etc.
    (ii) large families are no longer an economic benefit; children likely to survive; contraception available etc.
    (iii) completion of trends started in stages II and III.

ANSWERS TO PRACTICE QUESTIONS

Question 1 – Student's answer

Part (a) is a simple task if you have learnt the stages of the model, as outlined on p. 42. Make sure that you can refer to the birth rate, death rate and changes in the total population at each stage. Brief reference to the causes of the significant changes may well be needed to gain full marks.

    Part (b) contains a phrase which crops up many times in this syllabus – reference to *your chosen case study*. Other syllabuses might refer to more than one study. In the answer that follows the case study is France:

In the Middle Ages France's total population was probably about 20 million people, although it fluctuated, and probably fell to about half this figure during the Hundred Years War and the Black Death. This is fairly typical of a country in the *First Stage*, although the total was probably higher than in any other European country at that time.

    At about the start of the eighteenth century the population started to rise. At this time France was probably more economically, socially and scientifically advanced than any other country in the world. Advances in agriculture and commerce produced a strong economy, and allowed the country to support the increasing population. However, the country as a whole did not start to urbanise and industrialise to the same extent that Great Britain did as the century went on. Paris was the only city in France where industry developed on a large scale, and even here the industry was mainly craft based, producing specialist luxury goods for the court.

This meant that France's population development was rather different from Britain's at the same time. Agriculture was becoming more efficient, but there was no great demand from industry for a labour force, so although the death rate fell, as the model suggests it should, the birth rate did not stay high. Large families did not bring economic benefits as they did in more industrial countries. France's time in *Stage 2* was very limited, and she did not have the massive surge of population that other countries did.

In the late nineteenth century France did start to industrialise, and so might have expected a rapid growth in population. Instead the death rate continued to fall, but the birth rate fell with it, almost like a premature *Stage 3*. Unfortunately, in 1870 they were defeated in war by Germany, and lost many men and a large part of their territory. This loss of 'breeding stock' and the blow to national pride have both been given as reasons for the failure of the birth rate to stay high, despite industrialisation. Between 1851 and 1901 the population only grew by 10 per cent. Between 1901 and 1946 the growth was nil. The loss of young men, and national traumas of invasion in the two World Wars, meant that many women were left widowed and childless, and the country was in *Stage 4*.

In conclusion it can be said that France did not follow the 'classical' pattern of the model. Stage 2 and Stage 3 were combined and merged. This is partly because France's economy did not have the same type of industrial revolution that was experienced by other European countries. At the start they were more economically developed, and they followed a slower, more relaxed development pattern than Britain and Germany. It can also be seen that social factors may have been more important influences in France than they were elsewhere.

> **Examiner's note** This question created a problem for the candidate. His example did not fit the model very well, and the reasons for the deviations were not really economic ones. He made a very good job of finding some relationships between France and the model, and linking these to economic factors. Then he gave a valid explanation, worth very good marks, of why the deviations happened.

Question 2 – Student's answer

Question 2 cuts across two chapters of this book, which shows again how dangerous it can be to limit revision and ignore some sections of the syllabus. Geography is a subject about interrelationships. The issues raised in (b) are discussed in Chapter 7.

Part (a) has two command words, *outline* which asks for a fairly low level skill; and *discuss* which asks for high levels of skill. Extracts from a candidate's answer are given here:

Basically Malthus said that population, when unchecked, will increase at a geometrical ratio (1 – 2 – 4 – 8 – etc.) and double every 25 years. At best, food supply cannot be made to increase at other than a mathematical rate (1 – 2 – 3 – 4 – etc.). Disaster is inevitable, unless people 'restrain from marriage'...

> **Examiner's note** This is an uncomplicated idea, and does not need to be over elaborated. A simple statement like this is quite sufficient, for the first part of the question.

... writing in the early part of the nineteenth century Malthus could not have predicted the great advances in food production that would take place in the next two centuries. Developments included the agricultural revolution, the discovery and development of the prairies, etc., improved food packing and preserving, the growing use of chemicals ...

> **Examiner's note** Having outlined the developments, one or two of these should be discussed in detail, with examples of places, where possible.

... by the early 1960s it seemed as though his predictions were about to come true in the ELDCs, especially SE Asia, where geometrical population growth was taking place. The safety valve of emigration was not available here, and the Neo-Malthusians said famine was inevitable. Then the Green Revolution took place ...

> **Examiner's note** A whole new sequence can be developed here, giving location details and explaining the developments.

... and now, at last, population growth seems to be coming under control. A combination of improved living standards (which provide the motivation) and new kinds of contraception (which provides the means) is bringing the population explosion in the ELDCs under control. It finally appears that Malthus is about to be proved wrong!

> **Examiner's note** This is a very clear conclusion, which finishes the essay off well.

Question 3 – Student's answer

In order to answer part (a) it is important to 'define the terms' used in the question. This should be kept as brief as possible though:

Part of the difficulty with knowing whether optimum population has been obtained lies in the difficulty of defining what optimum population means. It clearly refers to the balance between population and resources, in a country or region. Overpopulation occurs when there are too many people to obtain a reasonable standard of living from the resources available in the country. Underpopulation is when there are not enough people to exploit the resources to their maximum efficiency. The idea of optimum population implies that the best possible balance between the two has been achieved.

> **Examiner's note** This is the only easy way to come anywhere near a definition. This candidate has done it clearly and logically.

However, the optimum population cannot be static. Changes are continually upsetting the delicate balance. These changes could include:

- *New technology, which makes resources more productive. The Green Revolution did this in India from about 1964 (although no one would suggest India had an optimum population).*
- *New resources are discovered, or become usable. This happened in the UK in the 1960s, when North Sea oil was discovered, increasing the resource base. This should have meant that the country could support more people.*
- *Old resources run out, or become uneconomic. This has led to the large scale unemployment in the Durham coalfield.*
- *Migration happens. In the 1960s Germany was short of workers. It was underpopulated and could not rebuild its industry fast enough. 'Guest workers' were brought in from countries like Turkey. They helped growth, but when the building had been done some people said that they were no longer wanted and were making the country overcrowded. What was the optimum population here?*
- *Social changes which are always affecting the population total. In the last 40 years the UK has had a low BR. This has not caused the population to drop, yet, but the population is aging and the dependency ratio will soon increase. The population is not changing, but the standard of living might start to fall. We could move from optimum to overpopulation. Then the old people will die, as they reach a certain age. The population will fall, the dependency ratio will fall, and the optimum could be restored again.*

All the above examples show that population and resources are in a dynamic, ever-changing relationship. The optimum may be achieved, but only for a short time.

> **Examiner's note** A general idea is stated. Then five examples of ways that this could work are given. All the examples are located in specific places, and two of them are worked through in detail. Then the final, brief paragraph ties everything together and draws a clear conclusion.

Part (b) is a very open-ended question. Answers will depend on the areas that have been studied by individual schools. However, here is how the candidate's answer was planned and structured:

Classification of policies			Example
Population policy	Increase population	Increase BR Increase immigration	France 1960s Germany 1960s; Israel
	Decrease population	Decrease BR – Persuasion – Compulsion Reduce immigration Encourage emigration	India pre state of emergency China 1 child policy UK 1970s onward UK 19th C (convicts)
Resource policy	Human resources	Develop through education	Asian 'Tiger' economies
	Physical resources	Seek new sources Conserve present resources Exploit more efficiently	Brazil – Amazonia UK home insulation policy N Sea oil improved tech.

Examiner's note As with many of these questions, a clear structure has to be found, breaking the answer down into a series of parts. This table shows ten separate ideas, and with 15 marks available candidates do not need to give a lot of detail about each of these ideas. Two or three examples developed in reasonable detail would gain a very good mark.

Question 4

For part (a) the sums should be done as follows (though it does not need to be laid out in such detail!):

		Formula	A	B
Dependent population	=	(per cent <15) + (per cent >64)	46 + 8 = 54	28 + 19 = 47
Non-dependent population	=	100 – (dependent population)	100 – 54 = 46	100 – 47 = 53
Dependency ratio	=	$\dfrac{\textit{Dependent population}}{\text{Non-dependent population}}$	$\dfrac{54}{46}$ = **1.17**	$\dfrac{47}{53}$ = **0.89**

In the example above the working has been written out for your benefit only. It is not necessary to write it out in the examination. However, you are strongly advised to show your working, even if you use a calculator. Even if your answer is wrong you may gain some credit for good working.

For part (b) you must be very careful that you only do what the question asks, and only write about the 15–39 age groups. It should be noted that the mark scheme published by the examination board has made an error on that point, so take care if you use the mark scheme! It is probably wise to mark the 15–39 section on the pyramids, although it is quite difficult to work out where this comes by counting sections of the graph. The best way is to measure the distance between 15 and 65 years, and then find half of that. You need to make sure that you follow both commands. You must *identify* differences between the two pyramids. Then you must *explain* those differences, as below:

Pyramid A shows that the 15–20 age cohort makes up about 6 per cent of the total for both genders. By 39 there has been a steep fall, to about 2 per cent. Pyramid B is much more even, with both sides at around 3 per cent throughout this age range. A is typical of an ELDC, in Stages 2/3 of the demographic transition, with a high DR caused by malnutrition and poor health care. B is an EMDC, in Stage 4, with better health and food supply. The slight bulge at 30–35 may result from a post-war baby boom.

The first two sentences pick out the key figures for A, and use them to make a comment about steepness. The third sentence makes two comparisons – the percentages at 15 and at 39, and the gradient throughout the period. The fourth sentence explains by referring to ELDCs, and the demographic transition, and then mentioning two causes of the high death rate. The fifth sentence makes contrasts about all these points, and the last sentence notes another feature of B, and suggests a possible explanation.

Before you start to answer part (c) of this question it is most important that you realise that parts (i) and (ii) both ask you to refer to the same country or region. Many candidates might feel that they could write a good part (i) with reference to China, and the results of the 'one child policy'. Unfortunately many of the texts which deal with this do not refer to ways of solving the problems; and so (ii) would have to rely on trying to work out an answer, which could become rather unrealistic if the candidate does not have a detailed knowledge of the country.

Here is an answer plan for both parts (i) and (ii) and detailing Kenya, a good example of an ELDC in Stage II of the demographic transition:

Problems	*Solutions*
1 Population explosion as this age group comes of child-bearing age.	Family planning programme. Publicity/trained health workers/free/subsidised pills etc. Desire to use FP brought about by knowledge that children will survive if fewer born.
2 Unemployment, and poverty, which would make pop. planning less effective.	Development, which will need education and investment. This should meet basic needs and should employ local workers, so that effects are spread throughout society. Jua Kali workshops in Nairobi are good e.g. Spread development with appropriate technology.

For part (d) the UK is probably the best country to choose as an example. Themes that could be discussed include:

► increasing dependency ratio
► increased burden on fewer working-age people
► growing number of single old people, mainly women
► changes in types of home needed
► more small homes, sheltered housing, care homes etc.
► need for alterations to existing homes
► changing labour needs – nurses, carers etc.
► increasing value of the 'grey pound' leading to off-peak holidays, daytime leisure activities, 'pensioners' night' shopping, etc.
► decreased mobility has important geographic consequences for urban structures, as aged find it more difficult to travel to out-of-town shops etc. May strengthen local community shopping etc.

Note that some of these have important spatial implications for geographers. If possible emphasise these, rather than making the answer very sociology based. Answers will attain the high level marks if they develop ideas and elaborate on them. A list like the one above would not gain top marks. Some points must be developed.

Question 5 – Student's answer

This question refers to Brazil, which is a specific area of study on this syllabus. Other syllabuses could have very similar questions which refer to another named country, or to 'a country of your choice'. The comments below could be adapted to fit any of these questions. Key phrases in the question include, 'reference to specific examples', 'internal migration', 'present distribution and density', and perhaps most difficult of all, 'assess the extent to which'.

Sketch maps are invaluable, both as an aid to learning before the examination, and as a way of presenting a lot of information quickly in the examination. Here is a perfect example of a place where such a sketch could be used. Draw a simple outline of the country, with areas of dense and sparse population named and shaded; mark arrows to show the main migration flows to be referred to; and you have a splendid aid which shows a lot of good, precise place knowledge. Here is one such map drawn by a student:

Figure A5.1

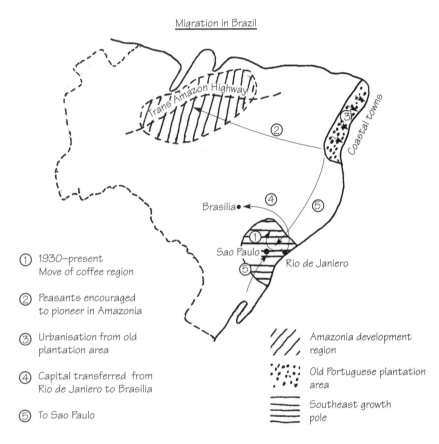

Migration in Brazil

① 1930–present
Move of coffee region

② Peasants encouraged
to pioneer in Amazonia

③ Urbanisation from old
plantation area

④ Capital transferred from
Rio de Janiero to Brasilia

⑤ To Sao Paulo

/// Amazonia development
region

Old Portuguese plantation
area

Southeast growth
pole

Five major population movements are shown on this map. They are:

▶ movement of the coffee-growing area from the 1930s to the present
▶ government encouraged migration from the northeast to Amazonia, in search of cheap land
▶ urbanisation from the overcrowded, rural northeast to nearby cities
▶ movement from the southeast to Brasilia
▶ movement from all over the country to Sao Paulo

This needs to be developed to explain how internal migration has influenced distribution and density. Then the answer must go on to *assess the extent* to which it has done this, if the answer is to attain the high level. Here is how one candidate did that:

Quite clearly internal migration has had an enormous influence on the present distribution and density, but two other factors need to be considered. These are immigration from outside Brazil, and natural population change. There has not been much immigration since the second world war. Small numbers of Europeans and Japanese have come into the country, but their numbers have been so small that they cannot be considered significant. Earlier than that, of course, immigration was the vital influence. Settlers from Europe were concentrated in the south and southeast, with slaves from Africa found mainly in the northeast, on the plantations.

When the plantation system collapsed it caused overpopulation to result, as the people no longer had sufficient resources. This triggered a lot of the migrations. It would be true to say that internal migration is adjusting the distribution and density of population that was originally established by immigration, but which had become unsupportable.

To what extent is natural change responsible for the present distribution and density? BR and DR must play a big part in the overall density, and the role of different BRs in different areas is very interesting. BRs tend to be higher in poor, rural areas than they are in urban areas, and if this was the only influence the rural population would be growing faster than the urban population. The reverse is actually happening because internal migration is more than making up for the difference.

So it can be concluded that, in the past 50 years, internal migration has been by far the most important influence on changes to Brazil's distribution and density of population.

> **Examiner's note** If you are asked to 'assess the extent to which ...' you must show an ability to weigh up different sets of things. This candidate gave good detail on internal migration, as did many others. Her answer reached a very high level because she went on to discuss other factors, and to reach a clear conclusion, based on and supported by evidence.

6 Solutions
Settlement

ANSWERS TO PRACTICE QUESTIONS

Question 1 – Student's answer
Part (a) deals with sources of evidence. With this board's examinations many questions ask about sources of evidence, and they usually ask for two separate sources. Be well prepared for this. Examples used for this question could include:

- ▶ old maps, from museum archives
- ▶ census returns, for the period since 1801
- ▶ parish records which go back before 1801
- ▶ trade directories, which give details of businesses at the time of writing
- ▶ archaeological evidence
- ▶ remaining old buildings, etc.

For the second part of (a) a student who chose census returns wrote:

Censuses were taken every 10 years (except 1941). They provide an accurate record of the number of people in each district of the country, and so they can be used to show how the town grew. They are very useful for studying towns that grew during the industrial revolution. The returns also show where people were born, and what jobs they did. The information could show the flows of people into the town, and also give a fairly good guide to the town's functions and how they changed as the town grew.

> **Examiner's note** This gives the simple, obvious use of raw numbers. Then it goes on to elaborate on two other uses for the data. The detail about industrial revolution towns is a nice, extra touch, which shows careful thought.

For part (b) students must use detailed local knowledge. Some sections of an answer written on Newcastle upon Tyne are given here:

... The wars, or skirmishes between the English and the Scots were partly economic, with both sides wanting plunder. They were also political, as the two rival groups fought for supremacy and control of land. The city became a garrison, both to protect its inhabitants against the raiders from Scotland, and to act as a base for army raids northwards.

The security that was offered by the castle and walls meant that the city became an important market. Markets for grain, cattle and other goods started inside the walls, and spread beyond them as the area became more peaceful and controlled. The Bigg Market and the Groat Market, can still be seen in the city to this day. Both are narrow streets, which broaden to provide an area where the market could be held. Both lie on the edge of the present CBD, and are narrow and winding, which proves difficult for car access. The Bigg Market is still the site of open-air trading, providing more access problems.

There are strong pressures to return the Bigg Market to its original use, and pedestrianise the street. It is clearly unsuitable for modern road traffic; its remnant mediaeval street pattern is still suited to its market function ...

... the port had made the town prosperous. Grainger, the politician and businessman, combined with Dobson, the planner and architect, to design and build some of the most elegant streets of houses and offices in the north of England. The street plan, and the facades of the buildings have been retained through all the periods of industrialisation and planning vandalism. The Grey Street area is still a model of town planning, and gives an air of dignity to the city. Linking with the mediaeval streets leading down to the Quayside this gives the city great potential for tourist development.

> **Examiner's note** This answer deals with all parts of the question. There are references to *political, social and economic processes;* the answer carefully limits itself to the *pre-industrial period;* but links are drawn with the *present townscape.*

Part (c) deals with issues. What is 'an issue' in geography? When you can say 'People are concerned about ... X ... , because ...' then you are talking about an issue. It is something which is a cause, or potential cause of conflict. If the issue has some sort of spatial element in it, then it is part of geography. In this question the potential issues, in Newcastle, and in many other cities are:

► jobs v. conservation
► jobs in tourism and 'heritage' v. jobs in industry
► beauty of environment v. convenience of modern living
► cars v. pedestrians
► profit v. quality of life, etc.

Each of these issues can be discussed in detail. One of the traps in answering this question is the temptation to discuss the issues in vague terms, and to forget the facts and the geography. The answer must always be linked to specific places, and how they should be managed. There will be at least two opinions about any topic. Candidates must present both opinions, and the facts that support those opinions. It is quite acceptable to decide in favour of one of the opinions, but full credit will only be given if the decision fits with the facts.

Question 2 – Student's answer

Here, a candidate's whole answer is presented without any interruptions. It shows how a straight essay question, without any guidance on structure, could be answered:

Orton and Tebay are villages in the east of Cumbria and I will use them to illustrate some national trends. They lie about three miles apart in the valley of the Lune, between the Lake District to the west, and the Pennines and Howgill Fells to the East. Tebay lies on the northern edge of the Tebay Gorge, a deep, steep-sided valley, cut when ice spilled across the watershed, cutting the gorge. The town was sited just above the flood plain of the Lune, on slate rocks. Orton lies north of Tebay, on a fairly flat area of limestones and shales. It is sited at a spring, and built on limestone, above the wet shale rocks.

Tebay is a linear settlement, which spread along the valley. Orton is nucleated, round a small market square, but it also has a more extensive open area, where fields on the shale are used for pasture. Tebay has about 500 people. Orton is a little bit smaller.

In 1960 Tebay was a railway village, with some farms and other services. Its position in the Tebay Gorge meant that it was an important coaching village on the main London to Glasgow route in earlier times. The main West Coast Railway was built through the same gorge in the nineteenth century, and there was a junction at Tebay, where a branch line was built eastwards. Four terraces with about 50 houses altogether were built to house railway workers, stretching the linear village still further along the valley.

Since 1960 changes in transport have continued to dominate Tebay's functions. The Beeching cuts in the mid-60s closed the branch line, and destroyed Tebay's main function, forcing its workforce to travel for railway work. When this work ended the village went into serious decline. Then in the 1970s the M6 was built, following the same route through the hills. This brought construction jobs, then service jobs, when transport cafes and service stations opened. Later still the natural route focus reasserted itself, and several transport company depots were set up in Tebay, trading with east Cumbria and along the M6.

In the 1980s the routeway was again chosen to build a major pipeline for North Sea gas. For over a year Tebay was the centre of construction, bringing many more temporary workers to the village. Now they have gone, and the village is trying to develop a tourist industry. Despite its old coaching inns and the trout fishing in the Lune, this has not been successful. With its dark slate houses, and railway terraces Tebay still looks like a working village. People stop off here, and spend some money, but do not stay. However, Tebay's cheap railway houses have become homes for people who commute to Kendal to work. These tend to be working-class people, not the usual middle-class commuters. They have been forced out of Kendal by rising house prices as Kendal is taken over by holiday makers and second home owners.

Orton's case is different. It remained a farming village until very recently. The market and water-powered mill had ceased long before 1960, so the village had few functions, and a steadily declining population. Some of the houses were taken over as second homes, but the owners contributed little to the village economy. After the M6 opened there was a gradual increase in commuting from the area. It is within easy reach of Lancaster and Preston, and just within commuting distance of Manchester. Incomers improved the appearance of the village by stripping the render and pebble dash off the houses, exposing the far more attractive limestone beneath. This led to a healthy building industry, as barns were converted and new houses built. Lying just outside the National Park meant there were fewer restrictions on building.

In the late 1970s the fashion for long-distance walks gave Orton another boost. The Coast to Coast Walk passes right through the village, and has brought many visitors. Several bed and breakfast establishments and cafes have opened, and the village shop, once on the verge of closure, has found a new prosperity. Farming, and related work is still Orton's main function, but the tourist industry has brought new prosperity and stemmed the decline.

These two villages illustrate what has happened to many communities in England since the 1960s. Tebay's functions have always been dominated by physical geography and the route which passes through. Though the means of transport keeps changing it is still a transport village, so there is continuity of function, in spite of technical changes. Orton exemplifies the change from farm-based villages to tourism and commuterism, typical of many English villages since 1960.

> **Examiner's note** Here, really detailed knowledge, gained through fieldwork, has been used. Two very contrasting villages are used, and several changes are discussed in some detail. The early paragraphs give detail on site and situation, which are not asked for, but which are needed for a full understanding of the answer. The last paragraph is excellent. It makes generalisations, which show how these two villages illustrate a national context.

Question 3

In order to answer this question a candidate must show knowledge and understanding of the three models mentioned in the question, and also of a chosen town. In addition the town's structure must be compared with the models. If the models have been learnt carefully it should be fairly easy to outline the main points of each (see p. 49–50). Some examples of notes on how towns could be compared with the models are outlined below. Thorough knowledge of the towns and the models would allow any one of these to be expanded into a full essay:

Sao Paulo – some of the models of development of cities in ELDCs have been based on Sao Paulo. It shows a ring structure similar to the Burgess model, but it is reversed, with the poorest housing on the outskirts. It has some of the features of the Hoyt model, though the industrial sectors have developed along the main roads, not railways as that model suggested they would.

continued

---

continued

Milton Keynes – this is a new town, and its development was planned. It did not grow and change over a long period, as Burgess suggested that cities do. It was planned so that industry was on the outskirts, separate from housing, so Hoyt's sectors did not develop. In some ways it can be described as a multiple nuclei town, because of the separation of functions, but even Harris and Ullman expected that development would happen over time, as towns grew out from their nuclei.

Bombay – the site of Bombay, built on a long peninsula, has had such a big effect on its growth that it does not fit any of the models very well. It has grown outwards, with new, spontaneous settlements on the outskirts, like Sao Paulo in some ways. It has industrial sectors along the river banks, like the Hoyt model. The growth of New Bombay could provide a new nucleus, so it could develop some of the features of the Harris and Ullman model.

Blackpool – some people have developed a model of a British resort town, based on Blackpool. In some ways it grew outwards, as Burgess suggested. It is special, though, because it has a CBD and also a Holiday Business District. The zones based on housing class are supplemented by zones based on class of holiday accommodation. The big hotels were built on the sea front, small hotels just behind the front, and boarding houses further back still. In recent years the retirement home function has become very important, again affecting the town's structure.

The Clydeside Conurbation – in many ways this was typical of the Hoyt model, with its industry spreading along the river, and attracting working class housing sectors. However, as Glasgow spread it absorbed other towns, often developed around other industries, each with its own zones of housing. This produced a type of multiple nucleus settlement. Since the 1960s redevelopment has radically altered the structure of the city. The old industrial sector at the heart has been largely replaced. Much of the working class now lives on the outskirts of the conurbation.

Question 4

'Social segregation' is when different classes, or income groups, live in different parts of a city; 'ethnic segregation' refers to different racial groups or nationalities. You need to write about both these segregations separately, although there may be some common points, and ethnic and class groups may overlap. In fact, you may even refer to Burgess's work on Chicago, where his ethnic quarters, like 'Little Italy' and 'Deutschland', became the basis of his class divisions.

'Economic factors' include things like the cost of housing, and the cost of travel to work. The cost of housing largely depends on the type of area it is in. People are willing to pay much more for more space, better environment, and even 'nicer neighbours'. This means that the desirable areas become more or less exclusive to rich people. There is also a tendency for the value of neighbourhoods to fall when poor people start to move in. There are many examples where the 'flight of the rich' has been observed. The concentration of both rich and poor in certain areas tends to be a process which gathers momentum.

In order to assess the extent to which segregation is *due* to economic factors you must also consider any other factors which play a part. Some possibilities are:

- politics and racism, as seen in apartheid South Africa and Nazi Germany, where ethnic divisions were enforced
- the desire for group support, as seen in British cities, where immigrant groups have deliberately concentrated in particular areas where their language is spoken, their religion is practised, etc.
- the two above factors could be combined, as people group together for support against what they feel is a hostile community

An example of a good concluding paragraph is given below:

Social segregation of income groups is often dependent on economic factors, because all groups have to live in housing they can afford, and only the rich can afford to live in the best areas. Ethnic segregation is sometimes based on other factors – but ethnicity is very often closely linked with economic status, or class. Immigrant groups are often poor, and so they are doubly restricted as to where they can live. In the USA some middle class Afro-Americans have moved out from the inner city areas, but for the majority of the group, poverty and colour combine to stop the move to the suburbs. The blacks have poor education, poor job prospects and poor housing. 'Other factors' do exist, but they often have a clear economic cause.

Question 5

There is a lot of information on the map, and on the key, for this question. Use it!

In part (a) you are asked for two reasons, but 4 marks are available; so you must elaborate each of your points. Two simple points are:

▶ land values differ as you move away from the centre
▶ the oldest buildings are found in the centre

These could be elaborated by:

▶ reference to the bid-rent theory and sorting of functions
▶ as transport changed the rich moved out, leaving poor houses for the poor people

For part (b) (i) only 2 marks are available, so the answer can be briefer:

Zone D has poor crowded housing, either Victorian terraces or council housing. Redevelopment may have brought high density flats. Zone A has bigger detached houses with gardens, garages etc.

For (b) (ii) there are 4 marks, so elaboration is needed:

Zone D is crowded because many people had to live as close to work as possible, in cheap mass-produced housing. When the old terraces were replaced cheap council or private housing was built in the same area. In Zone A the environment was less polluted and wealthy people were prepared to pay more for big houses in an exclusive area.

Part (c) refers to the age of industry in the inner and outer areas. As with any industry question you should consider referring to the different demands made by old and new industry for: space; raw materials; power; labour; markets; transport.

For part (d) you should have learnt the details of an example, so that you can draw a map quickly, accurately and clearly in an examination. You *must* be able to name specific areas and features in the learnt example. You can *annotate* the map or add a key, but note that there were no lines provided for writing here. You will not get marks for writing which is not clearly part of the map.

Question 6 – Student's answer

For part (a) there is quite a lot of data to deal with, and there are a lot of *differences* to be *explained*. These include differences in:

► % of poor in 1985 and in 2000
► absolute numbers of poor at both dates
► rates of change, of % and of totals, etc.

Very careful organisation is essential. Trends must be identified, and then explained. Examples must be quoted, to illustrate the explanations:

Both Asia and Africa expect the percentage of poor to increase by 50 per cent, and in Latin America it will increase by almost 50 per cent. In Asia and Latin America the numbers will also increase by about 50 per cent, but in Africa they will increase by over 100 per cent.

The rapid increase in all these three areas can be explained by:

● migration to the city, mainly of poor people, caused by:
   – pushes from the countryside   ⎫   Tutor's note: do not elaborate, this
   – pulls of the city   ⎭   comes in part (b) (i)
● possible counter-urbanisation or migration of the rich from the city
● natural population increase in the city, especially fast among the poor
● shortage of well-paid work in the city
● increasing concentration of wealth in the hands of few people, etc.

These trends are faster in Africa. African cities started their rapid growth later than the others, hence the rapid growth in numbers now. Civil conflicts are providing a strong push from the countryside, especially in East Africa, as is desertification in the Sahel. Instability and poor education levels mean that transnational development is not being attracted, as it is to Asia and Latin America (e.g. VW in Sao Paulo).

There are some counteracting trends in Asian and Latin American cities. For instance Singapore and Seoul have become very successful, and wealth is starting to 'trickle down' to the poor. Japanese cities have long had a high average income with few poor. Chinese policy has been to spread wealth evenly, again reducing the number living in poverty.

Meanwhile, in Europe/Mid East/N Africa (a strange grouping), the percentage living in poverty was high (2nd highest) in 1985, and is expected to fall to the lowest by 2000. Total numbers are rising though. The rapid growth of cities in Mid East and N Africa produces the growing total. Even in Europe the percentage of poor may not be falling as fast as might be expected, because in some of the most developed areas counter-urbanisation is leading to the rich moving out of the cities in increasing numbers, leaving a concentration of poor.

> **Examiner's note** This is a logical and thorough answer. The candidate has worked steadily through the data in the table. Many good points have been extracted from the information, and then discussed methodically. Well done! Note how the first paragraph has *used* the information. It has not just been lifted, but it has been manipulated to make a very good point. Also note how a reference to a model of development has been mentioned, in passing, in the fourth paragraph. This shows good geographical understanding.

For part (b) (i) the push–pull model gives an ideal structure for this question. Refer to a range of pushes, in different countries. Refer to a range of pulls, making reference to specific cities. Refer to conditions in EMDCs as well as ELDCs. The concentration of poor people in cities in the USA gives some parts of New York and other big cities an almost 'Third World' appearance.

For part (b) (ii) there is a list of suggestions for improvement of housing, employment, etc. in cities in ELDCs in Chapter 6 of this book. However, you could also include schemes to provide jobs and improve housing in inner city areas of cities in EMDCs. You could refer to Urban Development Corporations, EU funding, partnerships between private firms and local government, self-build schemes, and so on.

Do not just write a list, but on the other hand, do not write too much about one example. Try to give a number of case studies with a few developed points for each.

7 Solutions
Employment

Question 1 – Student's answer
The mark scheme for this answer made the following points:

> The question requires 'examining' and a mere listing would not reach pass standard. Some discussion of a limited range of practical measures – fertilisers, pesticides, genetic engineering – would gain a pass level mark.
>
> A middle level answer would show an appreciation of options of intensifying or bringing new land into production, supported by accurate and appropriate examples such as industrialised methods, irrigation schemes, etc.
>
> To gain the high level an answer would need, in addition, comment on more general influences – government policies, market forces, subsidies, etc.

This is a high level answer given by one candidate:

(a) A country can raise its agricultural output in several ways. More land can be brought into production, or production can be intensified on land that is already being used. The Dutch have increased their land area enormously during the twentieth century, by building dykes, around areas of shallow sea, like the Ijssel Meer. Then they pumped the water out, washed the salt out of the soil, added organic matter, and then farmed it. This is extremely fertile land, and it is farmed very intensively.

New land has also been brought into use in Israel. Large parts of the Negev Desert are now cultivated, using irrigation, and dry farming techniques to conserve water. This has allowed the population to grow as more refugees have returned to their homeland.

In the European Union great increases in output have resulted from the Common Agricultural Policy (CAP). This was an attempt to make Europe self-sufficient in agriculture. The CAP offered subsidies, and guaranteed prices to try to raise production. This allowed farmers to intensify by specialising in the best crop for the area. Because there was a guaranteed price for crops like wheat they knew they could afford to invest in fertilisers, pesticides, machinery and other capital goods. They were also encouraged to invest in improved, cross-bred strains of animals. Drugs, hormone treatment, controlled environments, and so on, all allowed animal productivity to increase. Greater efficiency, and small new areas of land could be produced by pulling up hedgerows, and so subsidies were offered to farmers to do this. No one considered the possible negative consequences at the time.

The Green Revolution in ELDCs like India was sponsored by the EMDCs, to try to enable these countries to avoid the famine that seemed imminent. Plant breeding was at the heart of the revolution. IR8 was the classic example of successful breeding. But the high outputs also needed high inputs of fertilisers, irrigation water and pesticides. All these had to be made available to farmers. This led to the development of industries. The chemical and seed multinationals, based in EMDCs, made huge profits from these developments – but the threatened famine was averted.

(b) The most important consequence of the improvements has been that the world's population has been fed. The massive famine that the Neo-Malthusians predicted has been avoided – so far at least. What is more, the standard of living of many people in the developed world, and some in the less developed world, has increased enormously during the last 50 years. But there has been a cost. The disaster may just have been postponed, not avoided altogether.

Examiner's note This is a dramatic introduction. It makes a very positive point, but provides a good lead in to the negative points which seem about to come. Good style like this puts an examiner in a positive frame of mind.

The most obvious point is that the increased output has allowed the population to go on increasing. This increase will inevitably lead to a limit being reached, as the 'limits to growth' theories say. The only question is, can people stay within the limits without reaching the disaster point? Even though the limits have not been reached worldwide, there are many small scale damaging consequences of increased production. I will deal with three main areas.

Examiner's note Candidates must choose a range of examples to get the full marks. Three thorough examples will be enough, as long as they deal with a variety of problems, at different scales and in different types of country.

The Sahel is an area of semi-desert. Population pressure caused intensification of the traditional slash and burn agriculture and nomadic pastoralism. Fallow periods became shorter, herds became bigger, and more wood was used for firewood. In the 1970s and 80s drought struck the area on a scale that had not been seen before. This may have been partly a natural event, but almost certainly it was made worse by human actions. Reduction of the vegetation cover over large areas of the region interrupted the water cycle, reducing interception and evapo-transpiration and increasing run-off.

Once the drought started it probably developed a feedback mechanism which made it worse. More and more people were pushed onto the fewer remaining well watered areas, and this in turn caused more disruption of the ecosystem. Loss of vegetation was followed by soil erosion, and desertification has resulted throughout the region.

Examiner's note This section deals with the problem in plenty of detail. The links between physical and human geography are important.

In much of the European Union, including East Anglia and the plateau of Beauce in France, cereal production has been intensified. This has led to monoculture over large areas. The wildlife has been almost wiped out by destruction of hedgerows, and use of pesticides. This has destroyed the food chain for the birds, and traces have also been found in sterile eggs. The widespread use of fertiliser has led to leaching of nitrates into streams, causing eutrophication. It is also contaminating water supply, leading to increased cost of purification.

The loss of hedgerows has had another consequence. They no longer provide windbreaks or roots to bind the soil, so a lot of valuable topsoil is being eroded away. Subsidies were paid to rip up the hedges; now more subsidies are being paid to replant them. The CAP has also been so 'successful' in increasing production that they now have to pay farmers to 'set aside' land and not grow crops on it. This is an enormous cost to the taxpayer.

Examiner's note In one area five or six problems are mentioned. Economic and environmental aspects are both dealt with. Again physical geography is well linked to human geography.

The final problem I will deal with is quite appalling. In order to intensify production of meat, farmers were encouraged to feed their animals on whatever was available. This led to the use of animal remains in feedstuffs. Animals that had died of diseases were ground up and fed back to their own species. Hens were even fed their own droppings to make sure nothing was wasted. As a result several diseases have spread through large parts of the British farm system. Edwina Curry said that all hens were infected with salmonella, and now it seems as though Mad Cow Disease has spread through the British cattle population. These diseases can spread to humans too. The saving on food has led to an incalculable cost in suffering for animals and humans, and an equally serious cost in slaughtered animals and lost sales.

These are some of the economic and environmental consequences of increases in the intensity of agriculture.

Examiner's note This is a passionate ending to the essay, and strong feelings are expressed. It is good to note that they develop from a well argued case, and are quite justifiable. The last sentence provides a very clear conclusion, which quotes from the title. It balances the positive opening very nicely.

Question 2

The factors which might be considered here are:

▶ raw materials
▶ labour supply
▶ transport

▶ energy
▶ markets
▶ government policy

If the changes over a period of time have to be discussed, then a fairly traditional industry which has undergone some interesting changes has to be chosen. The UK motor industry would be a good example. Three clear phases could be identified:

▶ early history of craft development
▶ concentration and mass production in the post-war period
▶ changes largely due to government policy and employment factors

However, the example below shows an essay plan, and sketch maps, for the US steel industry. Drawing sketch maps like these should be an important revision exercise.

The US steel industry shows gradual change from domination of raw material location to a flexible more market orientation. Improved transport and technology (especially efficiency of ore smelting) have allowed this.

▶ Stage 1 – early nineteenth century – 'fall line' sites for power – close to small iron ore resources – raw material based.
▶ Stage 2 – late nineteenth century – Pennsylvania coalfield – iron ore from Superior fields – Great Lakes as transport link – energy based.
▶ Stage 3 – early twentieth century – spread back along the Great Lakes route – developed at 'break of bulk' points (e.g. Cleveland). Also spread to Chicago-Gary area – Eastern Central coalfield + growth of market. 'Pittsburgh Plus' pricing system – 'stage managed industrial inertia' – political influence to protect the big investments of the steel companies – reduced spread of the industry (especially damaging to Birmingham Al).

Figure A7.1 USA – Steel industry in the 19th and early 20th century

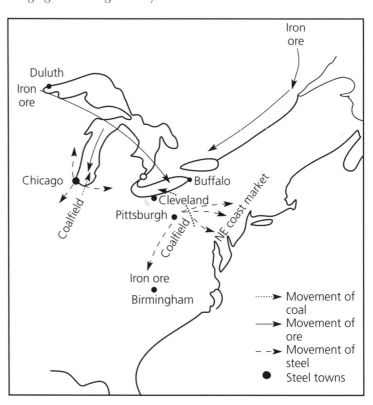

continued

| continued |

▶ Stage 4 – WW2 – spread to south and southwest – strategic factors – also development of strong market pull, especially in California.

▶ Stage 5 – post-WW2 – further decentralisation – growth of Eastern Seaboard plants (e.g. Sparrow's Point) – market influence now strong – improved transport technology – raw material sources more varied – strategic need to conserve US ore – market is also the source of scrap, which is now a major raw material.

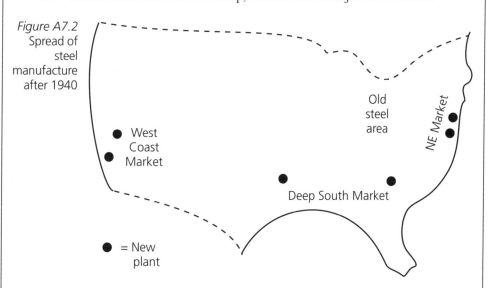

Figure A7.2
Spread of
steel
manufacture
after 1940

West Coast Market

Old steel area

NE Market

Deep South Market

● = New plant

Question 3

For part (a) (i) this question provides a lot of data. It is vital that you are used to dealing with such complex material in a logical and structured way. If figures are 'lifted' straight from the table and quoted in the answer they may gain some marks, but not many. 'Using' the figures by comparing, working out ratios and percentages, etc., is a higher level of skill, and will gain more marks. Here is an example of a high level answer:

In Region A primary employment fell drastically, from 55–10 per cent of the workforce. Some workers have gone into manufacturing, which increased by 50 per cent over this period. More have gone into services, which tripled its percentage share of the workforce. The LQs for primary and tertiary were both slightly above 1 in 1950. By 1990 the primary LQ was well below average, at 0.7, but secondary was well above, at 1.5. The 45 per cent of the workforce in tertiary matched the national figure, at an LQ of 1. Region B is less developed. The percentage in primary has fallen, but is still high, and services have risen, but are still low. Secondary has kept its percentage share, but its LQ has fallen, as the rest of the country developed manufacturing faster.

In both (i) and (ii) there is a strict limit on space. A concise writing style is essential. Despite a very compact style the answer below runs to five lines rather than the four allowed for part (ii). This is just acceptable, but any more would probably not be marked. Practice writing concisely!

This could be caused by decline in primary work, as farming is mechanised and workers are laid off. Industrial employment grew fairly slowly, maybe also a result of high capital investment. Growth of industry would need more workers in support services, like banking. Development, and rising living standards, might lead to more demand for personal services. Maybe many tertiary workers are underemployed.

For part (b) the location factors for manufacturing are energy, raw materials, labour, capital and market. Refer to these in your answer, for example:

> Many ELDCs lack coal, which was originally needed for industry. When they had raw materials these were often developed by, and exported to, EMDCs. Poverty means that the workforce is poorly educated, and markets are poorly developed. All this means that ELDCs lack capital to develop industry. When MNCs invest in ELDCs they often employ few workers, because plants are mechanised.

This is very concise and it manages to make relevant references to five factors affecting industrial location. The references are all relevant to explaining the answer. This type of answer would easily gain the full 3 marks available.

Part (c) gives an opportunity to refer to examples which has not been given in this question. The outline answer below concentrates on Bombay. It deals with both economic and environmental consequences thoroughly:

> Bombay is built at the end of a long peninsula, with its CBD at the tip, and its port along the east coast. Therefore the site is very congested, especially as industry spread west from the port, making access to the CBD very difficult. Housing, industry and transport all compete for land. This crowding and mix of functions causes environmental problems. Land prices in the CBD are very high, but shanty towns exist near the CBD. Oil refining and other heavy industries pollute the air and waste pollutes the sea. Concentration of industry brings agglomeration benefits, but it has created a 'core', and the inland 'periphery' suffers a 'backwash' effect.

The first sentence gives three relevant facts about the position and growth of Bombay. The second sentence brings in the idea of congestion, and elaborates it by showing how the port brought industry, and how this adds to congestion in a specific area. The third sentence elaborates further by listing functions. The answer then mentions the problems of high land prices, and the reverse – shanty towns. This combines economic and environmental consequences.

The sentence on pollution gives specific detail by referring to oil refining and heavy industry, and refers to air and sea pollution – making four linked points in one sentence. The final sentence shows good geographical understanding. Reference is made to two models. The idea of agglomeration economies comes from the Weber model, and the core/periphery model is referred to clearly. This sentence shows a nice balance between the advantages and disadvantages of the concentration on the coast, and rounds off a good answer very well.

Question 4

This is a very complex essay question. Very careful analysis of the question is essential before the answer is attempted. Here are some of the key words and phrases that could have been underlined or highlighted, and which should guide candidates in their answer:

► transnational corporations
► within and between countries
► information, investment and products
► effects of these changes
► organisation of business
► new technologies
► around the world
► attitudes ... to these effects

This plan shows how these key words could be used. Note that with such an open-ended and far-reaching question this list contains many suggestions – but many more could have been included.

Nature of the changes:
- ► easy and rapid movement of capital
- ► 'GNP' of many TNCs is greater than that of some ELDCs
- ► division of functions – to ensure maximum benefits to TNC, at the least cost:
  - – management in parent country
  - – research and design in very developed areas – educated labour – footloose, because information can then be transferred quickly – IT, Internet, etc.
  - – initial manufacture in fairly developed areas – skilled, quite cheap labour
  - – mass manufacture in ELDCs – cheap labour (e.g. computers Japan and USA to 'tiger economies' of Asia) (also e.g. trainers)
- ► factory locations in countries with good labour relations, government incentives, access to large markets (e.g. Nissan in Sunderland, 'new' car manufacture area – no old union practices – government aid, tax breaks, infrastructure investment, access to Europe both physically and within tariff system)
- ► compare the last point with Ford closing down Halewood on Merseyside
- ► sometimes 'dirty' jobs are exported, because ELDCs have less strict regulation of pollution and safety
- ► TNCs can bring political change; some of the TNCs with investment in S Africa brought pressure on government to end apartheid

Sources of evidence:
- ► figures for GNP from World Bank
- ► figures for TNCs from company reports
- ► details of company investments from their publicity releases

Effects of the changes and attitudes to them:
- ► governments want TNCs to bring investment and jobs (e.g. Brazil wanted to develop a motor industry)
- ► workers already employed may be suspicious – fear wage cuts
- ► unemployed welcome the chance of any jobs
- ► mobility of capital means that investment may not be long term
- ► cheap labour has become an even more important location factor
- ► growing importance of IT as an industry and source of power
- ► growth of international financial service sector (e.g. London, Hong Kong, Singapore)
- ► growth of inward investment in EU to have access to wider market; EU attempts to standardise labour laws, to equalise competition for jobs
- ► rapid growth of Asian Pacific Rim economies
- ► decline of heavy industry in Europe and USA

Question 5

The assessment objectives in geography syllabuses say that candidates should be tested on their Knowledge and Understanding. One aspect of this is an appreciation of people's values, and how they affect geographical decision making. Tourism is an area where conflicting values over the use of land and economic development are often in evidence.

This question looks for knowledge and understanding in (a) and reference should be made to case studies; (b) gives opportunities for showing understanding of fairly complex economic factors; and (c) is particularly concerned with understanding the influence of competing values in influencing development. Below are notes for an answer to (a) and (b), and a full answer to (c):

(a)

▶ Up to the 1950s tourism in UK based on mass transport by rail (and coach), therefore centralised in major resorts – Blackpool, Scarborough, Margate etc.

▶ Growth of private car ownership allowed decentralisation; spread to Cornish villages, Yorkshire Dales etc.

▶ Car access is damaging some areas (e.g. bans on cars proposed in parts of Peak Park; growth of holiday homes destroying traditional Cotswold villages).

▶ Car (and coach) transport allowed development of 'day visit' theme parks, like Alton Towers.

▶ Cheap air transport allowed package holidays – at first in Spain; gradual spread through Mediterranean, N Africa, etc.

▶ Bigger aircraft allow longer flights, and greater economies, so now many British tourists to Florida, US West Coast, Kenya, Seychelles etc.

▶ This needs investment in airports, roads, etc. May distort economies for short-term tourist development (e.g. Spanish resorts now trying to diversify their economies).

▶ This has had a damaging impact on British resorts. They have had to adapt – short breaks, specialist holidays (e.g. Southport Dance Weekenders).

(b) Positive impact – investment and high volume of tourist expenditure provides jobs for local people. Likely if country has:

▶ broad economic base, trained personnel, developed infrastructure
▶ some local investment in tourism to give local control
▶ policy to encourage reinvestment of profits
▶ conservation of environment to encourage long-term development

Negative impact – dividends sent out of country. High-paying jobs go to foreigners. Much of tourist revenue leaks out of country. Likely if:

▶ dominated by foreign investment
▶ economy has narrow base and cannot produce food, souvenirs, etc.
▶ unsophisticated and poorly educated population, unable to get jobs
▶ destroys previous economic base for short-term gain. Likely if:
  – environment is destroyed by pollution, destruction of wildlife, etc.
  – seasonal employment attracts workers, damaging previous economy
  – minimal government involvement or control
  – industry grows too quickly – overwhelms facilities

(c) What is 'cultural identity'? It is probably a sense of belonging to a particular group, with a set of values, beliefs and shared experiences. Some people in remote and isolated societies have developed cultural identities which do not owe anything to outsiders. The Indians of the Amazon rainforest have been isolated like this, and so they have a very special identity.

However, most groups of people have cultural identities which are the result of sharing ideas with their neighbours. When they travel and meet people they learn from their hosts, and hope that the hosts learn something too. In the early part of the century many British went on holiday in the Alps, and it was their influence that led the local people to develop their skiing and mountaineering industries. This was hardly 'eroding their cultural identity' as it allowed the development of a whole new identity combining Alpine traditions with a modern approach.

However, this was perhaps because the two groups were more or less evenly matched in terms of education, development and culture. It is quite different when a very rich and powerful group of people visit poor people in great numbers. Groups, like the Maasai of Tanzania, have been isolated for many years. Their culture is based on cattle herding, and a struggle for survival in a harsh environment. It should not surprise us if they are overwhelmed by visitors who come with powerful vehicles, build luxurious hotels, and treat the local people as quaint tourist attractions.

continued

> *continued*
>
> Luckily many of the Maasai have been able to adapt. Many of them have taken advantage of the developments and have become Park Rangers, guides and drivers for the tourists. Their traditional way of life has developed and modernised, but they have stayed on the land and still have contact with the old way of life. This is change, not erosion.
>
> Other groups have not been so lucky. In the resorts like Mombasa the tourists have taken over some areas completely. Land has been taken from the locals, who are forbidden from using the beaches and the new facilities in case they embarrass the tourists. Their way of life has been degraded, and they subsist by begging. Contact with the outside culture offers these people prostitution, Michael Jackson records, and McDonalds. This is a case of erosion of culture, but maybe urbanisation had changed the culture before the tourists arrived.
>
> Whether tourism leads to erosion or growth of culture depends on many things. The size of the tourist industry, the resilience of the original culture, the attitude of the government to development, and the level of pride of the people. Egypt has tried to develop its tourist industry. At one time this might have threatened to erode their Islamic values. Now it has led to the growth of fundamentalism, and attacks on the tourists to stop them polluting the culture. This is not an erosion of cultural identity, but a strengthening!

8   Solutions
World and regional development

SOLUTIONS TO REVISION ACTIVITIES

1 Answers are given within the text in Chapter 8.
2 The diagrams below show the answers.

Figure A8.1

(a) Flows from periphery to core:

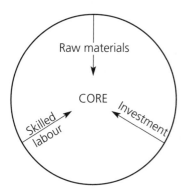

(b) Two alternative diagrams for reversal of the flows:

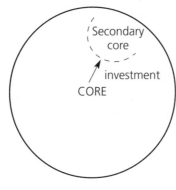
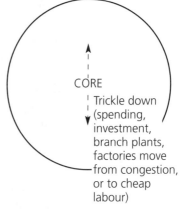

ANSWERS TO PRACTICE QUESTIONS

Question 1
A complete answer to part (a) is presented here.

> A 'newly industrialising country' is one of the countries which was in the Third World in the 1950s. That meant it had a low GDP and a very high proportion of its population engaged in agriculture. Since then it has invested heavily in the development of its manufacturing industry, so a large % of its population now work in that sector. In Rostow's terms it has achieved 'take-off' and is now on the 'drive to maturity'. Brazil and South Korea are good examples of NICs, although Brazil's industrialisation has not spread to the whole country yet.

This answer defines the phrase two ways – in terms of employment structures, and in terms of the model. These are both relevant. Two good examples are given to finish the question nicely.

The following answer to (b) (i) is an example of a very detailed answer to this part of the question:

> India is not very far on the way to being a NIC. It has several, isolated areas of manufacturing industry in what is basically still an agricultural country. The traditional core area of heavy industry is the Damodar Valley in the east, near Calcutta. Deposits of coal, and the Singhbum iron ore field, made this an ideal area for large scale investment. Railways, multi-purpose dams, and iron and steelworks were built, to support a series of steelworks. The biggest and newest is at Vishakapatman, on the coast to the south. The steel is used in shipbuilding here, and is also sold throughout the country.
>
> In the early Indian five-year plans this area soaked up a lot of the investment resources (which is typical of a core region in Myrdal's view). Other areas developed more slowly, and the industry was less capital intensive. Calcutta has its textile industry, as does Bombay and Bangalore. Factories here meet India's demands, and also export some cloth, but there is also textile manufacture in villages throughout the country, to meet local needs. This is small scale, low capital, labour intensive – much along the lines suggested by Ghandi. In Bombay there is also a fairly traditional footwear industry (described brilliantly in Seth's novel *A Suitable Boy*).
>
> Recently Indian industry has been modernising though. The two areas of major development are Bombay, with its oil imports providing the basis of a petrochemical industry; and the south, between Madras and Bangalore, where cotton machine engineering has developed into general engineering, with even aircraft manufacture.
>
> The Green Revolution increased India's demand for fertilisers and pesticides, and so many multinationals set up factories in India. This illustrates another aspect of the nature of India's industry. The plant at Bhopal was allowed to operate with less strict safety standards than similar plants in EMDCs. This led, inevitably, to disaster, killing and blinding many people. It just shows that, while industrialisation may be essential for India if they wish to develop the economy, it must be treated with care. Otherwise the advantages may be outweighed by the disadvantages.

Clear, precise facts need to be provided for part (b)(ii). To 'describe the impact' really means that you must describe change. Give details about the structure before industrialisation, and again, afterwards. Clearly the secondary sector will have increased; but how has this affected both primary and tertiary?

Question 2 – Student's answer

Regional development deals with ideas that are very similar to those in world development issues. Probably the UK is most often chosen as an example in regional development questions, but here the choice was Italy:

Italy's regional disparities are as wide as in any country in the Developed World. This meant that in the 1980s over one third of the EU's budget for regional development aid was going to Italy.

The pattern is comparatively straightforward – leading to a North/South division. The North is developed and industrialised, and forms the core. The South, or the Mezzogiorno, is agricultural and underdeveloped and forms the periphery. Rome is close to the dividing line, and falls just inside the prosperous North. The worst affected areas of the Mezzogiorno include the Naples region and Sicily. Sardinia is also clearly part of the periphery.

These areas which form the periphery can be distinguished by:

- higher unemployment than the North;
- a higher proportion of people employed in agriculture;
- a big out-migration of population throughout the post-war period;
- low levels of industrialisation;
- low wage rates, and a concentration on traditional industries (such as textile manufacture) even when industry has been attracted to the region;
- poor education and health care levels;
- traditionally low levels of investment in infrastructure (although this is being remedied with investment from central government).

Even in 1984, after much aid and also much out-migration, the 4 worst affected southern regions had just over 30% of Italy's population, but well over 50% of the unemployment. Even this, and the low wage rates that go with these unemployment rates, has not attracted enough industry to the area.

The core, on the other hand, has high levels of development. Industry is modern, technology-based and design-based, has grown rapidly. The great successes have been in motor manufacture, electrical and electronic goods and chemicals and pharmaceuticals. All these industries have enjoyed increasing sales, and have produced high profits allowing further investment. They have needed an educated and trained workforce, and so such trained people have been attracted in. Educational facilities in the area have also received plenty of investment.

As closer European integration has taken place the North of Italy has become part of what could be described as the 'European Core'. Easy access to the growth regions of southern France and southern Germany has been made possible by much investment in the trans-Alpine routes. This access to markets and influence has increased the strength of northern Italy – and also increased the peripheral nature of the South.

The reasons for the differences between core and periphery can be classed as geographical, historical and cultural, although they are all clearly linked.

Geographical influences include position in Europe, as discussed above. Climate, relief and soils also play a part. The climate of the South is considerably drier and hotter. The summer drought is more marked than in the North, and farmers do not have the benefit of rivers like the Po bringing irrigation water from the Alps. Relief in the South is often steep, with few areas of flat land for either intensive farming or factory and road construction.

This is in marked contrast with the flat Piedmont of the North, where the industrial triangle has developed, and where intensive agriculture has flourished. Soils in this area are much thicker and more fertile, thanks to deposition by the rivers flowing from the Alps. These rivers can also be used to produce HEP for northern cities; whilst the southern rivers are too small and unreliable to allow such development.

Historically too, the North has been at the heart of European developments. Since the Renaissance it has been an area of education and scientific development. The culture of the South has been locally based and inward looking. The strength of the church and traditional family bonds (extending to the development of the Mafia culture) have slowed development down. This isolation has been increased by the geographical shape of Italy – with the peninsula increasing in isolation as one moves south.

Despite the large amount of regional assistance given to the South since 1945 the core/periphery distinction remains. The best efforts of the Italian government and the EU cannot change the basic shape and geography of the country; and the recent emergence of strong regional parties in the North suggest that assistance may not continue to flow to the periphery in the same way.

> **Examiner's note** Each paragraph of this essay deals with a separate idea or group of ideas:
>
> 1 Shows the extent of the problem, and puts it in a wider context.
> 2 Defines the regions; a map would have helped.
> 3 Describes the problems of the periphery, listing a number of criteria.
> 4 Elaborates on the previous point, using some clear, useful statistics.
> 5 Describes the core – note that the points are developments from paragraph three, and are not just opposites.
> 6 Develops the points from 3, 4 and 5 in a wider context; it refers to the core periphery model, though.
> 7 Introduces the second part of the essay.
> 8 Describes 'geographical reasons for the development of the periphery'.
> 9 Contrasts the geography of the core; again it does not just give opposites, but develops some of the points in detail.
> 10 Ties the answer together well. It refers to historical developments, but then looks to the future. The rather pessimistic tone is briefly justified.

Question 3 – Student's answer

Part (a) shows a very detailed graph. At first glance it seems to provide too much information for what seems like a brief '4-mark opener'. You must note, though, that there is a lot of information here that could be used in later parts of the question, especially (c). Always treat questions as whole entities. Examiners often design them to 'flow' from one part to the next, with later sections building on what has gone in earlier parts. As for the two parts of (a), note that the first part asks about the bottom 'x' axis, and the second part about the vertical 'y' axis. Here is an answer to the whole question:

(a) Life expectancy (LE) is a good measure of economic development. As countries develop they can spend more on maternity care, health care and food distribution so people live longer. In 1965 LE in Indonesia was about 44 years, compared with 35 years in Guinea Bissau. A 25 per cent greater LE was clearly an indicator of more developed living standards.

Between 1965 and 1987 Indonesia's LE increased by 16 years, to 60 years. Guinea Bissau's increased by 4 years to 39. Indonesia's much faster development had led to a widening of the development gap and the Human Development Index.

> **Examiner's note** The candidate has selected the correct figures, and used them well. She has also, in her first three sentences, explained the relevance of the figures to the answer. Without this she could not have gained full marks.
>
> Note how she has used the abbreviation for life expectancy - at the first use she wrote it out in full, with the abbreviation. Subsequently she was quite justified in using the abbreviation.

(b) What is development for? If the aim was just to produce more wealth and more goods then economic criteria would be the only ones that were needed. If the aim was just to make the rich richer then the figures for social development would not matter, because the mass of the population would not matter. It would be all right for them to live in poor conditions. If the aim were just to make the country powerful and to impress the neighbours then they could point to a high GNP and say that they had achieved economic growth, and that is an end in itself.

I do not think that many people would accept that these are the aims of economic development. In fact the United Nations has introduced the Human Development Index (HDI) to try to give a balanced view of development in its member states, and to allow comparisons to be made between them. This HDI measures life expectancy, education and literacy, and income per person, and adds these together to give a single measurement. Because the measures of LE and literacy are average figures taken for the whole country, they show how well development is spread throughout the whole population. The figure is more balanced than GNP, because GNP could be very concentrated in a few people's hands.

From a purely economic point of view it may well be useful to measure social as well as economic development. Only a fit and educated workforce will be able to contribute to future development. A country that neglects its people also neglects its workforce, and they are needed to help growth in future. It is very noticeable that SE Asian economies, like Singapore, are putting a massive investment into education. This improves the quality of life for the people now, but it will also allow them to contribute to continuing development in future. This can be compared with countries like Zaire, where the limited wealth is largely concentrated in the hands of the president's family and friends. The ordinary people are ignored, and there is little hope of future development.

Maybe the UN will soon bring out some measure of how environmentally friendly development is. This will then allow people to measure whether development is really sustainable into the future.

> **Examiner's note** Three dramatic points are made at the start, with a good effect. Then the candidate goes on to give the simplest, and one of the best definitions of all round development. This is expressed clearly. The way that social development is linked with long-term economic development is clever. It shows real understanding of the topic, and gives a tight structure to the essay. The examples, though brief, are very relevant. The last paragraph takes the idea further, and though this was not specifically asked for, it is a very logical extension of the theme of the question. It showed that a very good candidate was thinking carefully about the topic.

(c) In Figure 1 there are many examples to illustrate countries that were probably still in Stage I in 1965. Sierra Leone, Guinea Bissau, Benin and Ethiopia all had a Life Expectancy of less than 45 years. This must indicate a high death rate, which probably was matched by a high birth rate which kept the population total more or less even. From 1965 to 1987 all those four countries saw an increase in LE of between 3 and 9 years. This suggests that they were starting to move from Stage I into Stage II, as their death rates started to be brought under control.

In 1965 India, Indonesia, Vietnam and China had progressed farther along the demographic transition road. They all had a longer LE of between 44 and 54 years. This suggests that in 1965 they had probably just entered Stage II. Over the next 12 years their LE increased to between 56 (India) and 69 (China). This is a remarkable change in a very short period of time. It suggests that a total revolution in health care and food supply had taken place. These countries were now at the peak of their Stage II growth.

It also suggests that, unless birth rate was also being brought under control, those countries would be experiencing a population explosion which threatened their survival. It was essential that they should progress beyond Stage II, and into Stage III and even IV, as soon as possible.

The cases of China and India make an interesting comparison. India is the world's biggest democracy and second most populated country. They were desperate to bring birth rate down and control the population, but as a democracy they had to attempt to do it by persuasion. They gave massive publicity to family planning and contraception, but tried not to compel people. They also realised that people would not accept family planning until they knew it was in their own interests, so they tried to develop the country, raise living standards, and give people security. All this took time, and on occasions the ideals were not followed, as in the period of the State of Emergency, when men were sterilised against their will.

In China the problem was more severe, because the population was higher. The answer could be more severe, because they had a communist dictatorship. The 'One Child Policy' was an attempt to bring the birth rate down, and to move through Stage III as quickly as possible. Incentives were offered to people to limit family size; but if they did not work there was a lot of compulsion. Group pressure was followed up by almost compulsory abortion and sterilisation if necessary. It now seems that China has gone through Stage III and the population, though still very high, is under control.

> **Examiner's note** This candidate made clever use of the data provided to describe Stage I, II and III, and the transitions between the stages. Then she went on to use her knowledge of India and China to become more specific in the second half of the essay. This produced a well balanced piece of work.

Question 4

Part (a) shows how linkages can operate in reverse to cause underdevelopment – a divider effect, rather than a multiplier effect. Many local examples could be chosen in the UK to illustrate the model, and one candidate's revision/essay plan notes are given below for one example. Note how many phrases are taken straight from the model and applied to the Merseyside context:

Merseyside port declined with the reduction of the Atlantic trade, and changing technology. This and other factors caused decline in shipbuilding. 'Linkages' in manufacturing sugar, milling, etc. were lost. Inner areas around Albert Dock became derelict. Environmental decline has only partly been halted by refurbishment. Heritage industry brought jobs, but fewer and less skilled than docks. Loss of rates revenue was one of the causes of the political crises and unrest in the 1970s and 80s. Militant Council. Toxteth riots. Many people left the city either for suburbs, or to other regions. New retail opportunities develop on the outskirts, e.g. along M62 and Warrington area. Continued decline in spending power in city. Further infrastructure decline keeps jobs away. Loss of morale and social decline. *The model seems to be very valid in this case!*

Note how the following plan for part (b) has many references to specific facts to be developed. It contains references to the models: Rostow; core/periphery; multiplier effect. If the plan has been learnt and understood these references are shorthand reminders which should stimulate detailed writing in the examination:

Brazil

(i)

- ▶ capital was available from successful coffee plantations
- ▶ this allowed development of infrastructure in SE region
- ▶ ports, railways, banking system created
- ▶ this needed steel industry
- ▶ raw materials available in Minas Gerais region
- ▶ market developed in SE region
- ▶ this created a clear core region, which attracted further investment
- ▶ industry stimulated by isolation from Europe during WW2
- ▶ import substitution needed, so industry started to take-off
- ▶ political decisions taken to encourage investment in 1950s and 60s
- ▶ military regime ensured population's compliance with plan
- ▶ especially wanted motor industry, as would have a strong multiplier effect to stimulate the drive to maturity
- ▶ this pushed the country through its take-off
- ▶ foreign capital for industry e.g. Ford, VW and Fiat – Brazilian capital for power stations etc.

(ii)

- ▶ rapid development of SE core Sao Paulo, Rio de J and Belo Horizonte
- ▶ mass migration from periphery, especially declining Northeast
- ▶ urbanisation – with all problems of the periphery (social and environmental), and congestion in the centre (economic and environmental problems)
- ▶ serious pollution in towns like Cubatao - coastal chemical industry
- ▶ damage to ecosystems and rural communities caused by HEP dams (e.g. Sao Francisco river scheme)
- ▶ massive international debt has led to strict monetary policy, wage cuts, widening gap between rich and poor, great demand for exports speeding the destruction of rainforest, etc.

Timed practice paper

PRACTICE PAPER 1

▶ The paper consists of eight questions

1 Rivers and river systems	**5** Population and resources
2 Ecosystems	**6** Human settlements
3 Meteorology	**7** Geography of employment
4 Climate	**8** Regions and regional development

▶ Answer any four questions.
▶ You have three hours to complete the paper.
▶ Candidates are encouraged to refer to their own fieldwork, practical work and detailed case studies where appropriate.
▶ You are reminded of the need for good English and clear presentation of the work.
▶ Credit will be given for maps or diagrams, where appropriate.

1 Rivers and river systems

(a) Study the sketch (Figure P1.1) which shows the work of the Tennessee Valley Authority (TVA) in the USA. The TVA has adopted an integrated approach to river basin management.

Figure P1.1

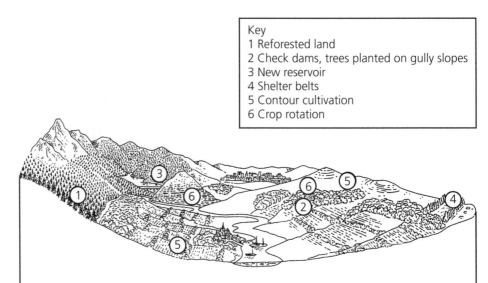

Key
1 Reforested land
2 Check dams, trees planted on gully slopes
3 New reservoir
4 Shelter belts
5 Contour cultivation
6 Crop rotation

Suggest reasons for *four* of the management strategies (1–6) shown on Figure P1.1. (12)

(b) With reference to a large river basin you have studied, explain how human activities can generate river management problems. (13)

(based on London)

2 Ecosystems

(a) With the aid of the diagram (Figure P1.2), explain and illustrate what is meant by:
  (i) ecological succession
  (ii) food chain (12)

(b) With reference to ecosystems that you have studied, explain how human activity can affect *either* ecological succession *or* food chains. (13)

(*London*)

3 Meteorology

Study Figure P1.3, a transect of an urban area and its rural fringe, showing climate data.

Figure P1.3

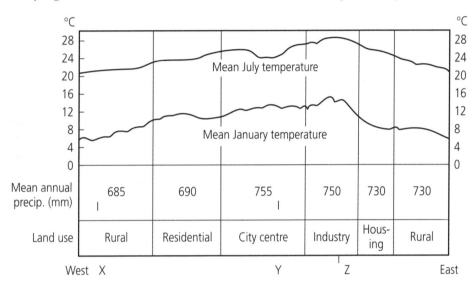

(a) (i) State the differences in temperature between the months of January and July for each of stations X and Y. [2 lines] (2)
  (ii) Explain *each* of the temperature differences given in your answer to (a) (i). [6 lines] (4)

(b) Suggest why the temperature peak at location Z is more pronounced in January than in July. [4 lines] (3)

(c) (i) What meteorological conditions would cause the greatest difference in temperature between the urban area and the rural fringe? [2 lines] (2)
  (ii) Explain your answer to (c) (i). [4 lines] (3)

(d) (i) Describe the variations in mean annual precipitation along the transect from west to east. [3 lines] (2)
  (ii) Suggest why these variations occur. [5 lines] (4)

(*London*)

4 Climatology

Explain and illustrate by reference to a region of continental or sub-continental scale, the cause and character of climatic seasonality in the tropics. (25)

(*Oxford*)

5 Population and resources

(a) Examine the causes of large-scale international migration. (12)

(b) For a country or region you have studied, discuss the social and economic consequences associated with such migration. (13)

(*London*)

Figure P1.2

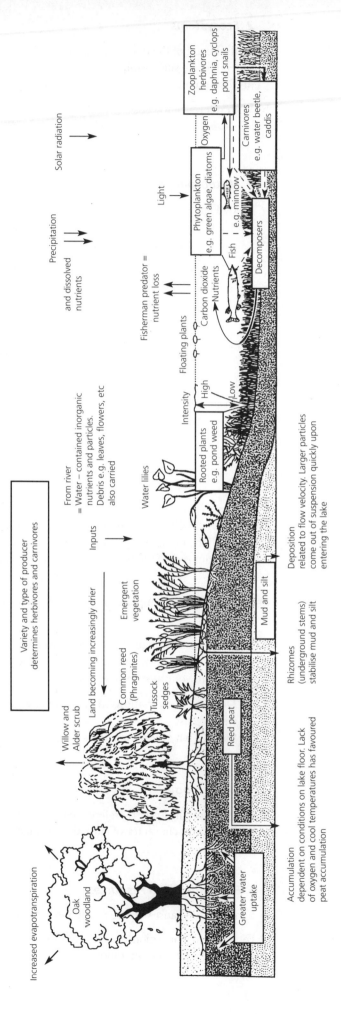

6 Human settlement

Either A

(a) What are the reasons for the continued growth of shanty towns (squatter settlements, spontaneous settlements) in and around many cities in the Developing World?   (10)

(b) In what ways are people and governments responding to the housing problems of these areas?   (15)

(AEB)

Or B

Discuss the role of Urban Development Corporations in urban regeneration in England. Illustrate your answer by reference to any *one* Urban Development Corporation.   (25)

(Oxford)

7 Geography of employment

(a) (i) Categorise the various types of service industry.   (4)

(ii) Explain why the distribution of people employed in service industry varies on a global scale.   (4)

(b) Study Table 1, which shows employment changes in a major city in a developed country from 1981 to 1987. Describe and offer an explanation for the changes.   (7)

Table 1

	Manufacturing Industry	Distribution, Catering, Repairs	Transport and Communications	Banking Insurance and Finance	Others
1981 (Total)	127 000	194 000	135 000	343 000	307 800
1981–4 (actual change)	–5 000	+6 300	–14 500	+31 200	+16 100
1981–4 (% change)	–4.0	+3.3	–10.7	+9.1	+5.2
1984-7 (actual change)	–43 000	–3 100	–17 200	+73 400	–12 600
1984–7 (% change)	–33.1	–1.5	–14.3	+19.6	–3.9
1987 (Total)	81 600	197 200	103 300	447 700	311 300

(c) With reference to an industrial region within a country, describe and account for the changes which have occurred in the nature and locations of industries within the region.   (10)

(UCLES)

8 Regions and regional development

(a) Identify *two* sources of evidence of links between 'First World' countries and 'Third World' countries and show how that evidence indicates the relationship between these types of countries.   (5)

(b) 'Africa, Asia and Latin America are no longer under colonial rule but are still dependent.' Discuss this statement by considering the relative importance of past colonialism and other factors in modern patterns of world trade.   (10)

(c) Describe and comment on the attitudes of government and other organisations towards the changing nature of trade links between First and Third World countries.   (10)

(NEAB)

ANSWERS AND MARK SCHEME FOR PRACTICE PAPER 1

1 Rivers and river systems

(a) Area 1 – prevent rapid run-off/increase infiltration/prevent soil erosion.
Area 2 – prevent soil erosion/increase infiltration/slow run-off/aid redevelopment of soil profile.
Area 3 – provide water supply/control and regularise run-off.
Area 4 – prevent flooding/aid navigation by scouring of bed/protect settlements/to cope with known maximum flows.
Area 5 – improve navigation/aid movement of flood water/dissipate floods during peak discharge.
Area 6 – aid navigation/keep peak discharge within banks/prevent build up of bars/stabilise the course.

Each strategy discussed gains 1 mark for a simple idea, which has not been explained well. 2 marks are awarded for one idea well explained, or several simple points not developed. Full marks given for one idea very well explained, with good use of geographical vocabulary, or for two or more points well explained.

(b) The answers will depend on the example chosen.
Low level marks will be achieved for answers which do not choose an appropriate case study and use it properly, e.g. small basin, or vague, unlocated work. Answers which lack structure, understanding or use of technical vocabulary also will get low marks.
Middle level marks are gained if there is some structure, understanding and explanation, but not complete. The management of the basin may not be dealt with thoroughly. High level answers are thorough, clearly structured, refer to detailed data from the example, and use good hydrological language.

2 Ecosystems

(a) The question asks for candidates to *explain* and to *illustrate*, with the aid of the diagram. A thorough explanation, using correct technical terms, with a detailed illustration making full use of the diagram, is needed for a high level mark on either (i) or (ii). There is a lot of information given, and it is not sufficient just to lift information; it must be used and developed.
If either the explanation or the illustration is not done thoroughly the answer is limited to the middle level of marks. It cannot gain more than 4 on either (i) or (ii). If only one of the commands is followed then the answer is unlikely to gain beyond the low level, unless it is done very well.

(b) There is a great variety of possible studies here. Note that the question refers to ecosystem**s** in the plural. This allows candidates scope to choose from a variety of examples to illustrate a variety of ideas. The examples should not be repetitive. Each one should be chosen to illustrate a different point, or set of points.
It is essential that good answers keep referring back to the case study being used to illustrate the point. Do not just mention the name of a place, and then write in purely theoretical terms. Make sure that the answer keeps being specific about the example. Write with a 'sense of place'.

3 Meteorology

(a) (i) X   16°C ± 0.5°
        Y   12°C ± 0.5°
    (ii) 2 marks for each:
        X   rural areas cool rapidly in winter – 1 mark
             2nd mark for explanation, e.g. no buildings to produce/store heat.
        Y   urban areas retain heat in winter – 1 mark
             2nd mark for explanation, e.g. buildings store and give off heat, reduce wind, etc.

(b) Z is a heat source in winter – 1 mark.
Award further marks for the ideas of industrial heating, more power station operation, larger factory buildings store more heat, etc.

(c) (i) High pressure/anticyclone – 1 mark.
Award further marks for elaboration, e.g. clear skies.

(ii) Anticyclones allow rapid radiation. Lack of wind to mix air and so reduces heating/cooling. Greater depth and range of answer allows access to the full marks.

(d) (i) Rises towards centre, then declines – 1 mark.
Rise and fall asymmetrical – 1 mark.

(ii) Rain-bearing winds from west – forced to rise by buildings – 2 marks for each basic idea.
Award the second mark in each case for elaboration on heat island, condensation nuclei, etc.

4 Climatology

Notes for the answer:

▶ Refer to movement of the overhead sun – movement of the area of maximum insolation, heat equator and ITCZ.
▶ Refer to the convective rainfall belt at the ITCZ.
▶ Describe the movement of the rainfall belt north and south of the equator.
▶ Lag between arrival of the overhead sun and ITCZ, and rainfall belt.
▶ Distinction between regions with rainfall at all seasons, two wet seasons, and short wet/long dry seasons.
▶ Precise locations and statistics for the different types of climate region needed.
▶ Effect of other factors on this pattern – land and sea (W Africa) – altitude (E Africa) – monsoon effect (S Asia).

The best answers to this question contain precise detail about specific places, usually located on a map. The general pattern is explained, but there is also some consideration of variations from that general pattern. Relevant technical terms are used to enhance the answer. The whole answer is developed clearly and logically. Explanation of the causes of seasonality, usually illustrated with simple, clearly labelled diagrams, is linked to precise details of its effects.

5 Population and resources

(a) In order to gain a basic pass mark it is necessary to give a relevant, named example of an international migration, and to describe at least two pushes and two pulls. If they are only listed and described, the answer will not get beyond the lowest pass mark.

As the discussion becomes more detailed, and the range of reasons becomes greater, the quality of the pass mark will rise. The high level of marks will only be achieved if candidates clearly link principles or models of migration to detail about the causes of the chosen case study. There must also be an acknowledgement of the complexity of causes. For instance answers may discuss how factors usually combine to cause a critical stress which makes people move; or the way in which facts causing migration become distorted by the filtering of human perceptions.

(b) Answers can deal with the consequences for the migrants, or the host population. The best answers usually consider both.

At least two key issues must be identified, and supported by evidence, if an answer is to gain a pass mark. However, if this remains more or less as a list, the mark will stay at a low level.

To achieve the middle level, detail is needed, such as the specific nature of the people who have migrated (age, gender, educational level, aspirations, etc.), and/or of the region they have migrated to (employment opportunities, growth rate, social composition of the host population, housing conditions and availability, etc.).

If the consequences of the migration are discussed in detail, a high level can be achieved. This could be expressed through consideration of the knock-on effects of migration on demography; structural imbalances caused by the migration; social issues, including housing, education and the impact on specific areas; or economic considerations including employment, cost of providing and administering resources; etc.

6 Human settlement

Option A

(a) Low level answers refer to the attractions of the cities, mainly in terms of work and 'bright lights', and to the absence of resources for providing 'proper' housing. They make few references to case studies, and the references lack detail and rely on generalisations. Note that credit cannot be given for the catalogue of negative features of shanty towns that are often found in low level answers; the question does not ask for this.

   Higher level answers provide precise details of the pull factors, particularly referring to jobs in industry, services and the informal sector. They discuss the problems of resources for housing, often making particular reference to competition for land and bid rent models. Mention is often made of the problem of finding land that is available for legal purchase and development. Shanty towns are seen as being the best available response to a difficult situation. Despite the problems they are seen as providing opportunities for people to establish themselves in the city.

(b) This question seeks detailed case study material. It asks for ways in the plural, and for reference to the contributions of both people and governments. At least two examples must be discussed for a low level pass to be achieved.

   High level responses refer to a variety of self-help schemes, site and service schemes, intermediate technology schemes, and so on. The roles of the individual and the local community are not neglected, with reference to the adaptability of shanty town housing to changing circumstances.

Option B

The first generation UDCs, in 1981, were London Docklands and Merseyside. The second group, in 1987, were Trafford Park, Black Country, Teesside, Tyne and Wear and Cardiff Bay; third group, 1988, were Central Manchester, Leeds, Sheffield and Bristol; fourth group, in 1992/3 were Birmingham Heartlands and Plymouth.

   Specific place knowledge must be shown. Links between urban problems and regeneration policies should be discussed. A pass level answer should show understanding of two problems, but for full marks at least four problems from the list below should be discussed:

   inner city unemployment
   population decline
   social imbalance and ethnic conflict
   environmental decay and derelict land
   factors deterring private investment
   crime and vandalism
   poor housing conditions
   problems of public health
   lack of access to open space
   land use conflicts
   accessibility and traffic management problems
   failure of local authority responses to urban problems
   local government financial restrictions, etc.

7 Geography of employment

(a) (i) Only 4 marks are available for this question, so an outline must be produced. Clarity of thought and organisation is more important than great detail.

   Categories could include:

retailing and trade	banking, finance, insurance, etc.
transport	leisure tourism, entertainment and sport
military	personal
administration	education
health	information handling, media, etc.[*]

*This could be subdivided from the rest, into quaternary services.
High level answers may recognise that there are many overlaps between any categories produced.

(ii) Again, only 4 marks are available. A mark should be awarded for recognising a distinction between the types of service employment available in more and less developed countries. Elaborations should mention the wide range of types of employment, levels of training and specialisation in the sector in some countries compared with much more limited employment in other countries. Distinctions between the large formal sector in EMDCs and the informal sector in many ELDCs could also be made.

(b) Low level answers note some of the changes, by lifting information from the diagram, and offer simple explanations for individual features of the change. Middle level answers start to order and manipulate the information and to see patterns in the information, but their explanations are not thorough or complete.

High level answers order the information clearly. They note that B, I and F was the largest employment sector in 1981, and that it grew throughout the period, with the rate of growth increasing. Meanwhile manufacturing and transport both fell throughout the period, with the rate of fall increasing. D, C and R, and Others rose slightly in the first part of the period, and then fell. Total employment increased by about 34 000 from 1981–4, but then fell by about 2 500 from 1984–7. High level answers then offer a reasonable explanation, based on economic cycles, geographic change, mechanisation and computerisation, or some such trend.

(c) There is enormous scope for answers here. Note that the location of the region chosen, the scale of the region, and the time span over which change has to be discussed, are all left to the candidate's discretion. Candidates should define how they intend to define the region and the period early on in the answer.

High level answers combine clear, precise and detailed knowledge of the chosen region (often with one or more maps), with intelligent, geographical analysis.

8 Regions and regional development

(a) There are many different indicators. Economic, social or political indicators could include:
► volume and nature of trade, indicating interdependence in terms of raw materials, manufactures and markets
► levels of investment
► levels of migration, to former colonial powers
► membership of the Commonwealth or French département status
► common legal/judicial system, etc.

(b) Many of the best answers start by discussing the nature of colonial trade. Then they go on to refer to ways in which the pattern of trade between former colonies and the colonisers has changed, and ways in which it has stayed the same. Trade in raw materials, traditional manufactured goods, preferential terms of trade, etc. might all be discussed.

'Other' factors might include:

► the rise of new trading relationships, for instance between former colonies, or with newly emerging powers like Japan, South Korea, OPEC and the former communist countries
► attempts to change the world trade system through GATT negotiations
► the development of NICs, such as Singapore
► transnational investments, etc.

(c) Governments of both the ELDCs and EMDCs should be considered, along with some examples of organised labour; World Bank; multinational corporations; development agencies; etc.

A variety of attitudes, from a variety of viewpoints must be described and explained. A recognition that these attitudes may be in conflict is essential for the achievement of high level marks.

LONGMAN
EXAM
PRACTICE
KITS

REVISION PLANNER

Getting Started
Begin on week 12

Use a calendar to put dates onto your planner and write in the dates of your exams. Fill in your targets for each day. Be realistic when setting the targets, and try your best to stick to them. If you miss a revision period, remember to re-schedule it for another time.

Get Familiar
Weeks 12 and 11

Identify the topics on your syllabuses. Get to know the format of the papers – time, number of questions, types of questions. Start reading through your class notes, coursework, etc.

Get Serious
Week 10

Complete reading through your notes – you should now have an overview of the whole syllabus. Choose 12 topics to study in greater depth for each subject. Allocate two topic areas for each subject for each of the next 6 weeks

No. of weeks before the exams	Date: Week commencing	MONDAY	TUESDAY
12			
11			
10			

Titles Available –

GCSE
Biology
Business Studies
Chemistry
English
French
Geography
German
Higher Maths
Information
Systems
Mathematics
Physics
Science

A-LEVEL
Biology
British and European
  Modern History
Business Studies
Chemistry
Economics
French
Geography
German
Mathematics
Physics
Psychology
Sociology

There are lots of ways to revise. It is important to find what works best for you. Here are some suggestions:

- try testing with a friend: testing each other can be fun!
- label or highlight sections of text and make a checklist of these items.
- learn to write summaries – these will be useful for revision later.
- try reading out loud to yourself.
- don't overdo it – the most effective continuous revision session is probably between forty and sixty minutes long.
- practise answering past exam papers and test yourself using the same amount of time as you will have on the actual day – this will help to make the exam itself less daunting.
- pace yourself, taking it step by step.

WEDNESDAY	THURSDAY	FRIDAY	SATURDAY	SUNDAY

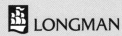 LONGMAN